2001 MARS ODYSSEY

2001 MARS ODYSSEY

Edited by

CHRISTOPHER T. RUSSELL
University of California, California, U.S.A.

Reprinted from *Space Science Reviews*, Volume 110, Nos. 1–2, 2004

KLUWER ACADEMIC PUBLISHERS
DORDRECHT / BOSTON / LONDON

A.C.I.P. Catalogue record for this book is available from the Library of Congress

ISBN: 1-4020-1696-4

Published by Kluwer Academic Publishers,
P.O. Box 990, 3300 AZ Dordrecht, The Netherlands

Sold and distributed in North, Central and South America
by Kluwer Academic Publishers,
101 Philip Drive, Norwell, MA 02061, U.S.A.

In all other countries, sold and distributed
by Kluwer Academic Publishers,
P.O. Box 322, 3300 AH Dordrecht, The Netherlands

Printed on acid-free paper

All Rights Reserved
©2004 Kluwer Academic Publishers
No part of the material protected by this copyright notice may be reproduced or
utilized in any form or by any means, electronic or mechanical,
including photocopying, recording or by any information storage and
retrieval system, without written permission from the copyright owner

Printed in the Netherlands

TABLE OF CONTENTS

Foreword   vii

R.S. SAUNDERS, R.E. ARVIDSON, G.D. BADHWAR, W.V. BOYNTON, P.R. CHRISTENSEN, F.A. CUCINOTTA, W.C. FELDMAN, R.G. GIBBS, C. KLOSS JR., M.R. LANDANO, R.A. MASE, G.W. MCSMITH, M.A. MEYER, I.G. MITROFANOV, G.D. PACE, J.J. PLAUT, W.P. SIDNEY, D.A. SPENCER, T.W. THOMPSON and C.J. ZEITLIN / 2001 Mars Odyssey Mission Summary   1

W.V. BOYNTON, W.C. FELDMAN, I.G. MITROFANOV, L.G. EVANS, R.C. REEDY, S.W. SQUYRES, R. STARR, J.I. TROMBKA, C. D'USTON, J.R. ARNOLD, P.A. ENGLERT, A.E. METZGER, H. WÄNKE, J. BRÜCKNER, D.M. DRAKE, C. SHINOHARA, C. FELLOWS, D.K. HAMARA, K. HARSHMAN, K. KERRY, C. TURNER, M. WARD, H. BARTHE, K.R. FULLER, S.A. STORMS, G.W. THORNTON, J.L. LONGMIRE, M.L. LITVAK and A.K. TON'CHEV / The Mars Odyssey Gamma-Ray Spectrometer Instrument Suite   37

P.R. CHRISTENSEN, B.M. JAKOSKY, H.H. KIEFFER, M.C. MALIN, H.Y. MCSWEEN, JR., K. NEALSON, G.L. MEHALL, S.H. SILVERMAN, S. FERRY, M. CAPLINGER and M. RAVINE / The Thermal Emission Imaging System (THEMIS) for the Mars 2001 Odyssey Mission   85

G.D. BADHWAR / Martian Radiation Environment Experiment (MARIE)   131

P.B. SAGANTI, F.A. CUCINOTTA, J.W. WILSON, L.C. SIMONSEN and C. ZEITLIN / Radiation Climate Map for Analyzing Risks to Astronauts on the Mars Surface from Galactic Cosmic Rays   143

In Memoriam, Gautam D. Badhwar (1940–2001)   157

FOREWORD

While not the brightest planet in the night sky, the red planet may be the most fascinating to the observer. Its redness draws one's attention. It says, 'I am different from the rest'. Indeed it is very different from the other bodies we see in the night sky and in many ways it is Earth-like. Moreover, since the discovery of the putative 'canals' on Mars, speculation has been rife (off and on) that Mars had once harbored life. Mariner 4 initially crushed those expectations by revealing an apparent Moon-like landscape. While 'first impressions' are supposedly lasting impressions, the higher resolution data from the later Mariner Mars missions, the Viking orbiters and landers, Mars Pathfinder, Mars Global Surveyor, and now the 2001 Mars Odyssey mission have completely changed that view. Once again speculation about life on Mars is on the increase.

The 2001 Mars Odyssey mission contributes greatly to the debate though its measurements of the neutron flux from near surface water, the detection of carbonates and its measurements of the radiation environment of the planet. This volume describes the Mars Odyssey and its payload. The introductory paper by R. S. Saunders and colleagues describes the mission, the spacecraft, and the early operations. The second paper by W. V. Boynton and colleagues describes the instrument that observes the neutrons and gamma rays coming from the surface giving us estimates of the amount of near-surface water and the elemental composition of the surface respectively. The next article by P. R. Christensen describes the thermal imager whose objective is to determine the mineralogy of the Martian surface. The following article by G. D. Badhwar describes the Mars Radiation Environment Experiment. While this article was in review, Gautam Badhwar passed away and F. A. Cucinotta graciously stepped in and took the article through the revision process. The next article by P. B. Saganti et al. discusses risks to future astronauts at Mars. The volume closes with an obituary for Gautam Badhwar.

The completion of this volume is due to the efforts of many individuals especially the referees and authors who worked together to produce what we believe is a most readable and complete description of the mission. We also wish to thank Anne McGlynn who assisted me during the initial assembling of the volume and Marjorie Sowmendran who took over after Anne retired.

C. T. Russell
University of California, Los Angeles
March 2003

2001 MARS ODYSSEY MISSION SUMMARY

R. S. SAUNDERS[1,6], R. E. ARVIDSON[2], G. D. BADHWAR[3], W. V. BOYNTON[4],
P. R. CHRISTENSEN[5], F. A. CUCINOTTA[3], W.C. FELDMAN[7], R. G. GIBBS[1],
C. KLOSS JR.[1], M. R. LANDANO[1], R. A. MASE[1], G.W. MCSMITH[1], M. A. MEYER[6],
I.G. MITROFANOV[8], G. D. PACE[1,11], J.J. PLAUT[1,*], W.P. SIDNEY[9],
D.A. SPENCER[1], T.W. THOMPSON[1] and C.J. ZEITLIN[10]

[1] *Jet Propulsion Laboratory, 4800 Oak Grove Drive, Pasadena, CA 91109-8099, U.S.A.*
[2] *Washington University, Department of Earth and Planetary Sciences, St. Louis, MO 63130, U.S.A.*
[3] *Johnson Space Center, NASA, Houston, TX 77058-3696, U.S.A.*
[4] *University of Arizona, Department of Planetary Sciences, Lunar and Planetary Laboratory, Tucson, AZ 85721, U.S.A.*
[5] *Arizona State University, Department of Geological Sciences, Tempe, AZ 85287-6305, U.S.A.*
[6] *NASA Headquarters, Washington, DC 20546-0001, U.S.A.*
[7] *Los Alamos National Laboratory, Los Alamos, NM 87545*
[8] *The Russian Aviation and Space Agency's Institute for Space Research (IKI), Laboratory of Space Gamma Ray Spectroscopy, Moscow, Russia*
[9] *Lockheed Martin Astronautics, Denver, CO 80201, U.S.A.*
[10] *National Space Biomedical Research Institute, Baylor College of Medicine, Houston, TX 77030, U.S.A.*
[11] *Now at Science Applications International Corporation, San Juan Capistrano, CA 92675, U.S.A.*
*Author for correspondence: Tel: (818) 393-3799; Fax: (818) 354-0966
Email Address: plaut@jpl.nasa.gov*

(Received 9 July 2002; Accepted in final form 5 March 2003)

Abstract. The 2001 Mars Odyssey spacecraft, now in orbit at Mars, will observe the Martian surface at infrared and visible wavelengths to determine surface mineralogy and morphology, acquire global gamma ray and neutron observations for a full Martian year, and study the Mars radiation environment from orbit. The science objectives of this mission are to: (1) globally map the elemental composition of the surface, (2) determine the abundance of hydrogen in the shallow subsurface, (3) acquire high spatial and spectral resolution images of the surface mineralogy, (4) provide information on the morphology of the surface, and (5) characterize the Martian near-space radiation environment as related to radiation-induced risk to human explorers.

To accomplish these objectives, the 2001 Mars Odyssey science payload includes a Gamma Ray Spectrometer (GRS), a multi-spectral Thermal Emission Imaging System (THEMIS), and a radiation detector, the Martian Radiation Environment Experiment (MARIE). THEMIS and MARIE are mounted on the spacecraft with THEMIS pointed at nadir. GRS is a suite of three instruments: a Gamma Subsystem (GSS), a Neutron Spectrometer (NS) and a High-Energy Neutron Detector (HEND). The HEND and NS instruments are mounted on the spacecraft body while the GSS is on a 6-m boom.

Some science data were collected during the cruise and aerobraking phases of the mission before the prime mission started. THEMIS acquired infrared and visible images of the Earth-Moon system and of the southern hemisphere of Mars. MARIE monitored the radiation environment during cruise. The GRS collected calibration data during cruise and aerobraking. Early GRS observations in Mars orbit indicated a hydrogen-rich layer in the upper meter of the subsurface in the Southern Hemisphere. Also, atmospheric densities, scale heights, temperatures, and pressures were observed by spacecraft accelerometers during aerobraking as the spacecraft skimmed the upper portions of the Martian atmosphere. This provided the first in-situ evidence of winter polar warming in the Mars upper atmosphere.

The prime mission for 2001 Mars Odyssey began in February 2002 and will continue until August 2004. During this prime mission, the 2001 Mars Odyssey spacecraft will also provide radio relays for the National Aeronautics and Space Administration (NASA) and European landers in early 2004. Science data from 2001 Mars Odyssey instruments will be provided to the science community via NASA's Planetary Data System (PDS). The first PDS release of Odyssey data was in October 2002; subsequent releases occur every 3 months.

1. Introduction

This paper provides an overview of the 2001 Mars Odyssey mission ('Odyssey'). This introduction provides the scientific context and objectives of the mission. Each of the science instruments is then described, and preliminary science results from the cruise and aerobraking phases are presented. Also included are a summary of mission operations, plans for science data archiving, and details on the spacecraft itself (Appendix A). This paper is intended to document the Odyssey mission prior to the main science data collection phase, which is now underway. Companion articles in this volume describe the science instruments in detail (Badhwar, 2004; Boynton et al., 2004; Christensen et al., 2004).

Odyssey is part of a long-term program of Mars exploration conducted by the National Aeronautics and Space Administration (NASA). The scientific objectives of this program are to: (1) search for evidence of past or present life, (2) understand the climate and volatile history of Mars, (3) determine the evolution of the surface and interior of Mars, and (4) prepare for human exploration (McCleese et al., 2001). The Mars Exploration Program is designed to be responsive to scientific discoveries. The guiding objective is to understand whether Mars was, is, or can be, a habitable world. To find out, we need to characterize the planet and understand how geologic, climatic, and other processes have worked to shape Mars and its environment over time.

Among our discoveries about Mars, the possible presence of liquid water, either in the ancient past or preserved in the subsurface today, stands out above all others. Water is critical to life, has likely altered the surface of Mars in the past, and is essential for future exploration. Thus, the common threads of Mars exploration objectives are to understand water on Mars, to identify past and present sources and sinks, and to understand the interaction and exchange between subsurface, surface, and atmospheric reservoirs, as well as the evolution of the volatile composition over time. To accomplish this, lander and/or orbiter spacecraft are launched at each Mars launch opportunity, approximately every 26 months. In 1997, NASA launched the Mars Global Surveyor (MGS), which together with the launch of the Discovery Program's Mars Pathfinder Lander, began a new era of Mars exploration. In the 1998–1999 launch opportunity, NASA launched the Mars Climate Orbiter (which failed to reach Mars orbit) and the Mars Polar Lander (which failed during its landing sequence). The 2001 Mars mission originally consisted of an orbiter and lander; both scheduled for launch in the spring of 2001. However, NASA decided to cancel the Mars 2001 lander and proceed only with the orbiter. This 2001 element of the Mars Exploration Program is focused on mapping the elemental and mineralogical composition of the surface, and monitoring the radiation environment (Saunders 2000, 2001a, 2001b; Saunders and Meyer, 2001; Saunders et al., 1999; Jakosky et al., 2001).

The 2001 Mars Odyssey mission contributes directly to the Mars Exploration Program goals by a direct search for water in the near surface of Mars at present and a search for evidence of past water in the surface mineralogy and morphology. In particular, 2001 Mars Odyssey carries instruments that will observe the Martian surface at infrared and visible wavelengths to determine surface mineralogy and morphology, provide global gamma ray and neutron observations for a full Martian year, and study the Mars radiation environment from orbit. The science objectives of the 2001 Mars Odyssey mission are to:

(1) globally map the elemental composition of the surface.
(2) determine the abundance of hydrogen in the shallow subsurface.
(3) acquire high spatial and spectral resolution images of the surface mineralogy.
(4) provide information on the morphology of the surface.
(5) characterize the Martian near-space radiation environment as related to radiation-induced risk to human explorers.

To accomplish these objectives, the science payload on 2001 Mars Odyssey consists of a gamma ray spectrometer, a multi-spectral thermal and visible imager, and a radiation detector as described in Table I. 2001 Mars Odyssey fills important niches in Mars exploration as a follow-on to the Mars Global Surveyor, and as a predecessor to the 2003 Mars Exploration Rovers (MERs), the 2003 European Mars Express orbiter, and the 2005 Mars Reconnaissance Orbiter (MRO). The Gamma Ray Spectrometer is a reflight of the instrument lost on Mars Observer. The Thermal Emission Imaging System (THEMIS) expands upon the results from MGS's Thermal Emission Spectrometer (TES) by examining the Martian surface in a similar spectral region, but with a resolution of 100 meters (30 times better than TES). THEMIS will also collect visible images at 18-meter resolution, bridging the gap between the few meter resolution of MGS's Mars Orbiter Camera (MOC) and the resolutions of many 10's to 100's of meters of the Viking Orbiter images. MARIE provides the first measurements of radiation in the Martian environment as a precursor to possible future human exploration of Mars.

Coordinated planning and implementation of science observations is provided by the Odyssey Project Science Group (PSG) that is comprised of the Principal Investigators (PIs), Instrument Team Leaders, and Interdisciplinary Scientists (IDS) (Table II). The PSG, which is chaired by the JPL Project Scientist and the NASA Program Scientist, establishes science policy for the project and adjudicates conflicts between instruments. Additional scientists (Odyssey Participating Scientists)

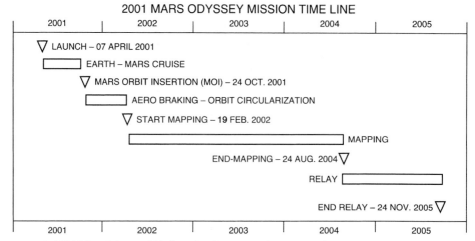

Figure 1. 2001 Mars Odyssey Mission Timeline. The primary mapping phase will be from February 2002 until August 2004, followed by a telecommunications relay phase until November 2005. Science observations in the relay phases beyond the mapping phase are currently undefined and depend on a number of factors, including spacecraft and instrument health as well as operational resources.

were competitively selected early in 2002 to assist with science operations and to augment the scientific expertise of the science teams. Data from the instruments will be distributed to the science community under the auspices of the IDS for data and archiving as well as the Odyssey Data Products Working Group (DPWG), a subgroup of the PSG. The DPWG has developed an archive plan that is compliant with Planetary Data System (PDS) standards and will oversee generation, validation, and delivery of integrated archives to the PDS. In addition, the 2001 Mars Odyssey mission will provide regular public release of images and other science and technology data via the Internet for public information purposes (see http://mars.jpl.nasa.gov/odyssey/, http://themis.asu.edu/, http://grs.lpl.arizona.edu/, http://marie.jsc.nasa.gov/, http://wwwpds.wustl.edu/missions/odyssey/; Klug and Christensen, 2001; Klug et al., 2002).

The 2001 Mars Odyssey mission timeline is shown in Figure 1. Odyssey's prime mission extends for 917 days from the start of mapping in February 2002 (following orbit circularization via aerobraking) to August 2004. During this prime science mission, Odyssey will also serve as a communications relay for U.S. and international landers in early 2004. After the prime science mission, Odyssey will continue to serve as a telecommunications asset for an additional 457 days. Thus, the total mission duration will be two Mars years (1,374 days). As a goal, an additional Mars year of relay operations is planned. Spacecraft resources may be available for extended mission science observations during the later relay phases of the mission. Observations beyond the prime mission could provide data on interannual phenomena, as well as global high resolution visible imaging and improved elemental concentration maps.

TABLE I
2001 Mars Odyssey Instruments

Instrument	Description	Principal Investigator
THEMIS (Thermal Emission Imaging System)	Will determine the mineralogy of the Martian surface using multispectral, thermal-infrared mages that have 9 spectral bands between 6.5 and 14.5 μm. Will also acquire visible-light images with 18-m pixel resolution in either monochrome or color.	Philip Christensen, Arizona State University
GRS (Gamma Ray Spectrometer)	Will perform full-planet mapping of elemental abundance, at a spatial resolution of about 600 km, by remote gamma ray spectroscopy, and full-planet mapping of the hydrogen (with depth of water inferred) and CO_2 abundances by combined gamma ray and neutron spectroscopy.	William Boynton, University of Arizona (GRS Team Leader) William Feldman, Los Alamos National Laboratory (Team Member for Neutron Spectrometer) Igor Mitrofanov Institute of Space Research (PI, High Energy Neutron Detector)
MARIE (Martian Radiation Environment Experiment)	Will measure the accumulated absorbed dose and tissue dose rate as a function of time, determine the radiation quality factor, determine the energy deposition spectrum from 0.1 keV/μm to 40 keV/μm, and separate the contribution of protons and of high charge and energy (HZE) particles to these quantities.	Gautam Badhwar, Johnson Space Center (PI, deceased) Cary Zeitlin, PI National Space Biomedical Research Institute

2. Odyssey Science Instruments

The 2001 Mars Odyssey science payload (as previously noted) consists of a Thermal Emission Imaging System (THEMIS), a Gamma Ray Spectrometer instrument suite (GRS), and a Martian Radiation Environment Experiment (MARIE). The GRS instrument suite includes the Gamma Subsystem (GSS), the Neutron Spectrometer (NS), and the High-Energy Neutron Detector (HEND). The locations of these instruments on the spacecraft are shown in Figure 2. Note that the GRS

TABLE II
2001 Mars Odyssey Project Science Group (PSG) Members

Raymond Arvidson (IDS for Data and Archiving)
Washington University
St. Louis, Missouri

Gautam Badhwar (MARIE PI, deceased)
NASA Johnson Space Center
Houston, Texas

William Boynton (GRS Team Leader)
University of Arizona
Tucson, Arizona

Philip Christensen (THEMIS PI)
Arizona State University
Tempe, Arizona

Cary Zeitlin (MARIE PI)
National Space Biomedical Research Institute
Houston, Texas

William Feldman (GRS Team Member for NS)
Los Alamos National Laboratory
Los Alamos, New Mexico

Bruce Jakosky (Interdisciplinary Scientist)
University of Colorado, Boulder
Boulder, Colorado

Michael Meyer (Program Scientist)
NASA Headquarters
Washington, DC

Igor Mitrofanov (HEND PI)
Institute for Space Research (IKI)
Moscow, RUSSIA

Jeffrey Plaut (Project Scientist after October 2002)
Jet Propulsion Laboratory
Pasadena, California

R. Stephen Saunders (Project Scientist until October 2002)
Jet Propulsion Laboratory
Pasadena, California

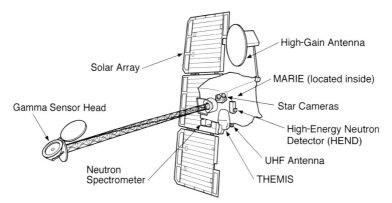

Figure 2. 2001 Mars Odyssey Spacecraft in Mapping Configuration. The Odyssey spacecraft as described in Appendix A consists of a central spacecraft bus with a 6-m boom for the gamma sensor head, and articulated solar arrays and high-gain antenna. Electrical power is provided by solar arrays. Communication to and from Earth is provided by low- and medium-gain antennas mounted on the spacecraft bus and an articulated high-gain antenna pointed toward Earth. Science instruments onboard the Odyssey spacecraft include a Thermal Emission Imaging System (THEMIS), a Gamma Ray Spectrometer (GRS) instrument suite, and a Martian Radiation Environment Experiment (MARIE). The GRS instrument suite includes the Gamma Subsystem (GSS), the Neutron Spectrometer (NS), and the High-Energy Neutron Detector (HEND). All of these instruments, except the gamma sensor head, are located on the spacecraft bus, with THEMIS pointed at nadir.

gamma sensor head is located on a 6-m boom, which is perpendicular to the orbit plane. All other instruments are located on the spacecraft bus, with THEMIS pointed at nadir. MARIE has a 68-degree field of view, which is pointed in the anti-velocity direction. Its field-of-view does not include Mars. The NS and HEND observe neutrons in all directions. A detailed description of the 2001 Mars Odyssey spacecraft is given in Appendix A.

2.1. THERMAL EMISSION IMAGING SYSTEM (THEMIS)

THEMIS addresses the 2001 Mars Odyssey objectives of acquiring high spatial and spectral resolution images of surface mineralogy, and providing information on the morphology of the Martian surface. These mineralogical and morphological measurements will help to determine a geologic record of past liquid environments and will help map future potential Mars landing sites. Specific THEMIS science objectives are to:
(1) determine the mineralogy and petrology of localized deposits associated with hydrothermal or sub-aqueous environments, and to identify sample return sites likely to represent these environments.
(2) provide a direct link to the global hyper-spectral mineral mapping from the Mars Global Surveyor (MGS) Thermal Emission Spectrometer by utilizing the same infrared spectral region at high (100 m) spatial resolution.

TABLE III

Thermal Emission Imaging System (THEMIS) Science Team Members

Team Member	Title	Organization
Philip Christensen	Principal Investigator	Arizona State University
Bruce Jakosky	Co-Investigator	University of Colorado, Boulder
Hugh Kieffer	Co-Investigator	U.S. Geological Survey, Flagstaff
Michael Malin	Co-Investigator	Malin Space Science Systems
Harry McSween	Co-Investigator	University of Tennessee
Kenneth Nealson	Co-Investigator	University of Southern California
Anton Ivanov	Participating Scientist	Jet Propulsion Laboratory
Melissa Lane	Participating Scientist	Planetary Science Institute
Alfred McEwen	Participating Scientist	University of Arizona
Mark Richardson	Participating Scientist	California Institute of Technology
James Bell	Participating Scientist	Cornell University
Greg Mehall	THEMIS Mission Manager	Arizona State University
Steven Silverman	Project Engineer	Santa Barbara Remote Sensing

(3) study small-scale geologic processes and landing site characteristics using morphologic and thermo-physical properties.
(4) search for pre-dawn thermal anomalies associated with active subsurface hydrothermal systems.

The THEMIS Principal Investigator is Philip Christensen of Arizona State University. THEMIS team members are given in Table III. Further description of THEMIS is given later in this issue (Christensen *et al.*, 2003).

To accomplish the THEMIS objectives, this instrument will determine surface mineralogy using multi-spectral thermal-infrared images in 9 spectral bands from 6.5–14.5 μm with 100-m pixel resolution. THEMIS will also acquire visible images at 18 m/pixel in up to 5 spectral bands for morphology studies and landing site selection. The THEMIS thermal-infrared spectral region contains the fundamental vibrational absorption bands that provide the most diagnostic information on mineral composition, as all geologic materials, including carbonates, hydrothermal silica, sulfates, phosphates, hydroxides, silicates, and oxides have strong absorptions in the 6.5–14.5 μm region. Thus, silica and carbonates, which are key diagnostic minerals in thermal spring deposits, can be readily identified using THEMIS thermal-infrared spectra. Remote sensing studies of terrestrial surfaces, together with laboratory measurements, have demonstrated that 9 spectral bands are sufficient to detect minerals at abundances of 5–10%. The use of long wavelength infrared data has additional advantages over shorter-wavelength visible and near-infrared data because it can penetrate further through atmospheric dust

TABLE IV

Thermal Emission Imaging System (THEMIS) Mapping Observations

Mineral mapping (daytime IR) during clearest periods and warmest time of day	∼50% of total mission data
Nighttime temperature mapping throughout the year	∼15% of total mission data
Visible imaging during late afternoon periods when shadows are better and IR data of lower quality	∼30% of total mission data
Color imaging for targets of opportunity	∼5% of total mission data

and surface coatings with absorption bands that are linearly proportional to mineral abundance, even at very fine grain sizes.

THEMIS was designed as the follow-on to the Mars Global Surveyor Thermal Emission Spectrometer (TES), which produced a hyper-spectral (286-band) mineral map of the entire planet. THEMIS covers much of the same wavelength region as TES. Furthermore, the THEMIS filters were optimized utilizing knowledge of Martian surface minerals determined from TES data, and TES global maps will allow targeting of areas with known concentrations of key minerals. THEMIS will achieve infrared signal-to-noise ratios of 33 to 100 for surface temperatures (235 –265 K) typical for Odyssey's late afternoon (about 3:00 to 6:00 p.m.) orbit. In addition, Odyssey's orbit is ideally suited to the search for pre-dawn temperature anomalies associated with active hydrothermal systems, if they exist. The visible imager will have a signal-to-noise ratio greater than 100 for Odyssey's late afternoon orbit. The THEMIS instrument weighs 10.7 kg, is 28 cm wide × 30 cm high × 31 cm long, and consumes an orbital average power of 5.1 W.

The THEMIS observation strategy given in Table IV shows that about half of the THEMIS data will be devoted to daytime, infrared mapping for minerals. The first few months of THEMIS operation, as shown in Table V, were devoted primarily to acquiring data for selection of MER landing sites and other targets of interest to the public and the scientific community.

2.2. GAMMA RAY SPECTROMETER (GRS) SUITE

The GRS instrument suite consists of three instruments: a Gamma Subsystem (GSS), a Neutron Spectrometer (NS), and a High-Energy Neutron Detector (HEND). These instruments address the 2001 Mars Odyssey objectives of globally mapping the elemental composition of the surface, and of determining the abundance of hydrogen in the shallow subsurface. Thus, this instrument suite plays a lead role in determining the elemental makeup of the Martian surface.

When exposed to cosmic rays, chemical elements in the Martian near-subsurface (in uppermost meter) emit gamma rays with distinct energy levels. By measuring gamma rays coming from the Martian surface, it is possible to calculate surface

TABLE V
Thermal Emission Imaging System (THEMIS) Early Observation Priorities

MER Landing Sites	Day/Night IR; VIS; whenever possible
	Meridiani
	Melas
	Isidis
	Eos
	Gusev Crater
Viking 1, 2, & Pathfinder Sites	Day/night IR; VIS; whenever possible
Valles Marineris layered deposits	Day/night IR; VIS; 6 observations
Southern hemisphere young gullies	Day/night IR; VIS; 6 observations
Putative shorelines	Day/night IR; VIS; 6 observations
Polar caps	Day/night IR; VIS; 6 observations
Geometric calibration sites	Day/night IR; VIS; 4/day
IR drift calibration sequences	Day/night IR; 3/day; Equatorial, polar, night
General interest	Outflow channels, southern hemisphere dunes, Arsia, Pavonis, Ascraeus, Olympus, Elysium Mons, polar dunes, Elysium flows, Day/night IR; VIS; 1/orbit

elements' distributions and abundances. In addition, measuring both gamma rays and neutrons provides a measurement of hydrogen abundance in the upper meter of subsurface, which in turn allows inferences about the presence of near-surface water.

The GRS objective is to determine the composition of Mars' surface by full-planet mapping of elemental abundance with an accuracy of 10% or better and a spatial resolution of about 600 km by remote gamma ray spectrometry, and full-planet mapping of the hydrogen abundance (with depth of water inferred) and seasonal CO_2 frost thickness. The GRS team leader is William Boynton of the University of Arizona. The GRS team shown in Table VI will operate this suite of instruments from the University of Arizona. Further description of the GRS instrument suite is given later in this issue (Boynton, *et al.* 2003).

The GRS will also address astrophysical problems such as gamma ray bursts, the extragalactic background and solar processes by measuring gamma ray and particle fluxes from non-Martian sources. For example, GRS data from extragalactic gamma ray bursts (GRBs) will be used with the data from Ulysses and near-Earth satellites (High Energy Transient Explorer-2, Wind, etc.). Interplanetary triangulation, a technique involving accurate timing of burst arrival times, allows the sky positions of the sources of GRBs to be determined with accuracy of several minutes of arc. GRS data can also provide insight into solar flares. The simultaneous measurement of gamma rays and high-energy neutrons from powerful solar flares at

TABLE VI
Gamma Ray Spectrometer (GRS) Team Members

Team Member	Title	Organization
William Boynton	Team Leader	University of Arizona
Igor Mitrofanov	HEND PI	Space Research Institute, Russia Academy of Sciences
William Feldman	Team Member for NS	Los Alamos National Laboratory (LANL)
James Arnold	Team Member	University of California, San Diego
Claude d'Uston	Team Member	CESR, Toulouse
Peter Englert	Team Member	University of Miami
Albert Metzger	Team Member	Jet Propulsion Laboratory
Robert Reedy	Team Member	LANL/University of New Mexico
Steven Squyres	Team Member	Cornell University
Jacob Trombka	Team Member	Goddard Space Flight Center
Heinrich Wänke	Team Member	Max-Planck-Institut für Chemie
Johannes Brückner	Team Member	Max-Planck-Institut für Chemie
Darrell Drake	Team Member	Techsource, Inc.
Larry Evans	Team Member	Computer Sciences Corporation
John Laros	Team Member	University of Arizona
Richard Starr	Team Member	Catholic University
Kevin Hurley	Participating Scientist	University of California, Berkeley
Thomas Prettyman	Participating Scientist	Los Alamos National Laboratory
G. Jeffrey Taylor	Participating Scientist	University of Hawaii, Manoa

Mars, combined with those from the vicinity of Earth, allows a stereoscopic image of the active region on the Sun. These stereoscopic observations of powerful flares will provide a three-dimensional model of the sources of hard-electromagnetic radiation and corpuscular emission in active regions on the Sun.

The GRS (as noted above) consists of several components. The gamma sensor head is separated from the rest of the spacecraft by a 6-m (20-ft) boom, which was extended after Odyssey entered its mapping orbit. This minimizes the interference from gamma rays coming from the spacecraft itself. An initial GRS calibration was performed during the first 100 days of mapping. Then the boom was deployed, and it will remain in this position for the rest of the mission. The NS and the HEND components of the GRS are mounted on the main spacecraft structure and will operate continuously throughout the mission. The entire GRS instrument suite weighs 30.2 kg and uses 32 W of power. The GSS measures 46.8 cm long, 53.4 cm tall, and 60.4 cm wide. The NS is 17.3 cm long, 14.4 cm tall, and 31.4 cm wide. The

HEND measures 30.3 cm long, 24.8 cm tall, and 24.2 cm wide. The instrument's central electronics box is 28.1 cm long, 24.3 cm tall, and 23.4 cm wide.

2.3. GAMMA SUBSYSTEM (GSS)

The GSS will detect and count gamma rays emitted from the Martian surface. By associating the energy of gamma rays with known nuclear transitions and by measuring the number of gamma rays emitted from a given portion of the Martian surface, it is possible to determine surface elemental abundances and discern their spatial distribution. While the energy in these emissions determines which elements are present, the intensity of the spectrum reveals the elements' concentrations. These energies will be collected with 600-km resolution over time and will be used to build up a full-planet map of elemental abundances and their distributions. The GSS uses a high-purity germanium detector cooled below 100 K to measure gamma ray flux. GSS performance is a strong function of its temperature, which in turn constrains the spacecraft orbit beta angle (angle between orbit normal and direction to Sun) to insure that the GSS cooler is shaded from the Sun. Thus, the orbit beta angle must be less than $-57.5°$ ($-56°$ to shade the cooler and a pointing uncertainty of $1.5°$) in order to acquire useful GSS data. Furthermore, annealing of the germanium detector on the GSS may be required to recover performance, with each annealing cycle taking approximately 10 days.

2.4. NEUTRON SPECTROMETER (NS)

The NS measures neutrons liberated from the near-subsurface of Mars by cosmic rays. Since Mars has a thin atmosphere and no global magnetic field, cosmic rays pass unhindered through the atmosphere and interact with the surface. Cosmic ray bombardment of nuclei of subsurface material down to about 3 m produces secondary neutrons. These neutrons in turn propagate through the subsurface and interact on their way out with subsurface nuclei. Fast neutrons produced by the cosmic rays may in turn be moderated by collisions with nuclei before they escape from the subsurface, resulting in neutrons with thermal or epithermal energies. The flux of secondary neutrons from the surface, as a function of energy, provides information primarily on the concentration of H and C in the uppermost meter of the surface material.

The NS sensor consists of a cubical block of borated plastic scintillator that is segmented into four equal volume prisms. In the mapping orbit, one of the prisms faces forward into the spacecraft velocity vector, one faces backward, one faces down toward Mars, and one faces upward. Neutrons coming directly from Mars will be separated from those coming from the spacecraft using a combination of velocity filtration (because the spacecraft in orbit about Mars travels faster than a thermal neutron) and shielding of one prism by the other three. Fast neutrons are separated from thermal and epithermal neutrons electronically. Details of the instrument and the Doppler filter technique for separating thermal and epithermal neutrons are described by Feldman *et al.*, 2001. The NS is provided and operated by the Los Alamos National Laboratory (LANL). William Feldman at LANL is the NS team leader within the GRS Team.

2.5. HIGH-ENERGY NEUTRON DETECTOR (HEND)

The HEND complements the NS as it measures the higher energy neutrons liberated from the Martian surface by cosmic rays. HEND consists of a set of five particle sensors and their electronics boards. The sensors include three proportional counters and a scintillation block with two scintillators. The proportional counters and an internal scintillator detect neutrons with different energies. With these sensors, HEND is able to measure neutrons over a broad energy range from 0.4 eV up to 10.0 MeV. HEND also helps calibration of the Gamma Subsystem. HEND is provided and operated by the laboratory of Space Gamma Ray Spectroscopy at the Russian Aviation and Space Agency's Institute for Space Research (IKI) in Moscow, Russia. Igor Mitrofanov is the Principal Investigator.

2.6. MARTIAN RADIATION ENVIRONMENT EXPERIMENT (MARIE)

MARIE addresses the 2001 Mars Odyssey objective of characterizing the Martian near-space radiation environment as related to radiation-induced risk to human explorers. As space radiation presents a serious hazard to crews of interplanetary missions, MARIE's goal is to measure radiation doses that would be experienced by future astronauts and determine possible effects of Martian radiation on human beings. Hazardous space radiation comes from two sources: energetic particles from the Sun and galactic cosmic rays from beyond our solar system. Both kinds of radiation can trigger cancer and damage the central nervous system. A spectrometer inside MARIE measures the total energy from these radiation sources; both in interplanetary cruise from Earth to Mars and in orbit at Mars. As the spacecraft orbits Mars, the spectrometer sweeps through the sky and measures the radiation field coming from different directions. Specifically, MARIE goals are to:
(1) characterize specific aspects of the Martian near-space radiation environment.
(2) characterize the surface radiation environment as related to radiation-induced risk to human exploration.
(3) determine and model effects of the atmosphere on radiation doses at the surface.

The Principal Investigator for the MARIE experiment originally selected by NASA was Gautam Badhwar, who unfortunately died just before Odyssey reached Mars (this issue, Cucinotta, 2003). The MARIE Principal Investigator is now Cary Zeitlin of the National Space Biomedical Research Institute, Baylor College of Medicine in Houston. The MARIE team members are shown in Table VII. Further description of MARIE is provided later in this issue (Badhwar, 2003).

The MARIE instrument was provided by NASA's Human Exploration and Development of Space (HEDS) Program in order to characterize the radiation envir-

TABLE VII
Martian Radiation Environment Experiment (MARIE) Team Members

Team Member	Title	Organization
Cary Zeitlin	Principal Investigator	National Space Biomedical Research Institute
Francis Cucinotta	Co-Investigator	Johnson Space Center
Kerry Lee	Co-Investigator	University of Houston
Timothy Cleghorn	Co-Investigator	Johnson Space Center
Ronald Turner	Participating Scientist	ANSER Corporation

onment at Mars. The instrument, with a square field of view 68 degrees on a side, is designed to continuously collect data during Odyssey's cruise from Earth to Mars and in Mars orbit. It can store large amounts of data for downlink whenever possible, and will operate throughout the entire science mission. The instrument weighs 3.3 kg and uses 7 W of power. It measures 29.4 cm long, 23.2 cm tall, and 10.8 cm wide.

3. Odyssey Operations: Launch, Interplanetary Cruise, Aerobraking, and Early Mapping

The 2001 Mars Odyssey spacecraft was launched from Cape Canaveral on April 7, 2001 at 11:02 a.m., EDT, during its first launch opportunity. One hour after launch, Odyssey's signals were received at the Deep Space Network (DSN) complex near Canberra, Australia. Odyssey's cruise from Earth to Mars was accomplished in 200 days, on a Type 1 trajectory taking less than 180° around the Sun. Further details of spacecraft activities from launch through cruise, arrival and aerobraking are given in Appendix B. Cruise navigation was done with two-way Doppler and ranging data as well as a series of delta differential one-way ranges (ΔDORs). These ΔDOR measurements established Odyssey's position in the plane-of-the-sky complementing line-of-sight ranging provided by the two-way Doppler and ranging data. There were four trajectory course maneuvers (TCMs); the first and second at 46 and 86 days after launch; the third and fourth at 37 and 13 days before arrival. The Odyssey spacecraft arrived at Mars on October 24, 2001 within 1 km of its aim point. The initial orbit period was 18.6 hours, well within the expected range of 15–24 hours. Then aerobraking was used to transition from this initial elliptical orbit to the desired near circular mapping orbit.

Aerobraking was conducted in three phases: walk-in, main phase, and walk-out. Walk-in, which was initiated 4 days after arrival and accomplished in eight orbits, lowered the periapsis altitude to 110 km. Then during the main phase, small

TABLE VIII

2001 Mars Odyssey Mapping Orbit Parameters

Epoch: (SCET–UTC)	12-SEP-2002 05:15:27
Index Altitude	392 km
Semi-Major Axis	3785 km
Inclination	93.2 deg
Eccentricity	0.0115
Orbit Period	1.964 hr
Longitude of Ascending Node	173.7 deg
Argument of Periapsis	268 deg
LTST of Descending Node (hh:mm:ss)	16:21:33
LMST of Descending Node (hh:mm:ss)	16:16:05
Solar Beta Angle	−59.0 deg

thruster firings at apoapsis kept the drag pass periapsis altitude to heights where atmospheric heating and drag on the spacecraft were within limits. The transition from aerobraking to a mapping orbit (the walk-out phase) was done with three maneuvers in mid-to-late January 2002. In total, aerobraking was done without incident during 330 orbits in 75 days.

At the end of aerobraking, the Odyssey spacecraft was in its near circular science-mapping orbit of 370 to 432 km, some 18 km above that of MGS. The science orbit as shown in Table VIII has an inclination of 93.1 degrees, which results in a nearly Sun-synchronous orbit. The orbit period is just under two hours. Successive ground tracks are separated in longitude by approximately 29.5 degrees and the entire ground track nearly repeats every 30 days. After achieving this science orbit, the final major spacecraft event before the start of mapping was the deployment of the High-Gain Antenna (HGA), which was successfully performed on February 4, 2002.

3.1. START OF MAPPING AND EXPECTED ORBIT EVOLUTION

The science-mapping mission began on February 19, 2002, 118 days after arrival. THEMIS was turned on and began imaging. A day later, GRS was turned on. The GRS's gamma sensor head had accumulated radiation damage during cruise. This necessitated a 'warm anneal' process, where it was warmed up and allowed to cool back to its normal operating temperature. The GRS warm anneal was completed on March 22, 2002 and improved the instrument performance. Although the GRS bands are somewhat broader than at launch, they were within specification. The GRS boom deployment occurred in early June 2002. Another in-orbit GRS anneal was performed in early November 2002.

A major success story at the start of mapping was the recovery of the MARIE instrument. MARIE experienced an apparent loss of 'heartbeat' in August 2002, two months before arrival. At that time, attempts to revive MARIE were unsuccessful. Once the mapping orbit was achieved, more extensive troubleshooting of the MARIE instrument began. Memory dumps were performed to determine the last states of the instrument. After that, MARIE's 'heartbeat' was re-established on March 6, 2002, and MARIE started returning science data from Mars orbit. The instrument has continued to return science data since then, and its prognosis for a long life is good.

The science orbit design is optimized to provide a balance between THEMIS and GRS observations; MARIE investigations do not directly affect the orbit design. At the start of the mapping orbit, the local true solar time (LTST) of approximately 3:15 p.m. allows high-quality THEMIS observations, but the beta angle of −44 degrees affects GRS data quality. High quality THEMIS infrared data are only obtained at LTST earlier than 5:00 p.m. while high quality GRS data are only obtained for beta angles less than −57.5 degrees. The first THEMIS opportunity started at the beginning of mapping and will continue for about 300 days. The second THEMIS opportunity starts at about 550 days and continues until the end of the mapping phase. THEMIS and GRS observations may continue outside of their optimal solar geometry, as power and telecommunication constraints allow.

The time-history of LTST and beta angle are controlled by changing the spacecraft orbit nodal precession rate (the rate at which the orbit plane rotates in inertial space). Figure 3 shows the expected time history of LTST and beta angle for the mapping and relay phases of the orbiter mission. The figure also includes local mean solar time (LMST), Mars-to-Earth range, and Mars-centered solar longitude L_S. LMST is a fictitious solar time that assumes that Mars moves in a circular orbit about the Sun with a period equal to the actual elliptical Mars orbit. LMST is constant for a Sun-synchronous orbit; differences between LTST and LMST are due to Mars' orbital eccentricity.

The nodal precession rate is controlled by slight changes to orbit inclination. The inclination of the science orbit (see Table VIII) is biased slightly higher than that required for Sun-synchronous precession to cause the beta angle to decrease (go more negative) so that GRS observations can commence early in the science mission. During the first 670 days of the mapping phase, the LMST drifts at a constant rate from its initial value of 3:54 p.m. to 5:00 p.m. At that time a maneuver using 8 m/s of delta-V will lock LMST to 5 p.m.

The Mars-to-Earth range is the dominant factor in determining the return data rate. This range is near the maximum during the first THEMIS opportunity, but is near the minimum at the beginning of the second THEMIS opportunity. Thus, more THEMIS data will be transmitted to Earth during the second THEMIS opportunity. Quality GRS data acquisition commences 154 days into the mapping phase and continues until the end of the mapping phase. During this GRS observation period, which spans more than one Mars year, the maximum beta angle is −54.5 degrees. MARIE will operate throughout the entire mapping phase of 917 days.

4. Odyssey Science Observations in Interplanetary Cruise, Aerobraking, and Early Mapping

4.1. SCIENCE OBSERVATIONS DURING INTERPLANETARY CRUISE

Odyssey science instrument operations during interplanetary cruise provided an opportunity to operate and calibrate the instruments. THEMIS obtained an Earth-Moon image that was the first to show the Earth and Moon in a single frame at nearly their true separation (i.e., perpendicular to the Earth-Moon line). MARIE operated from the beginning of interplanetary cruise until mid August 2001, when MARIE failed to respond to commands. This anomaly terminated MARIE observations until after orbit insertion and aerobraking were completed. In the four months MARIE was operating, it met its cruise science goals and two large solar particle events were observed. The interplanetary radiation environment, as a function of time and distance from the Sun, was measured and agreed with models (Cucinotta *et al.*, 2002; Vuong *et al.*, 2002; Saganti *et al.*, 2003, this issue).

All three of the GRS instruments were operated and obtained useful data. The GSS was operated during cruise and obtained nearly 1000 hours of data. The spectra from these cruise GSS observations will be used to remove the spacecraft signature from the gamma ray spectra at Mars. The NS was also operated during cruise and performed as expected (Feldman *et al.*, 2001). This showed that the flux of fast neutrons generated by cosmic-ray interactions with spacecraft material is about 1/3 of that expected from Mars. The HEND was operated for a large fraction of the cruise phase and detected many cosmic, solar, and soft-gamma ray events. From May 5 through September 24, 2001, HEND detected 25 gamma ray bursts (GRBs), which were confirmed by other spacecraft observations. Also, HEND detected six powerful solar flares, which were also confirmed by other spacecraft. The most intense solar flare event took place September 24, 2001 in the last four hours of HEND operations during interplanetary cruise.

4.2. SCIENCE OBSERVATIONS DURING AEROBRAKING

Odyssey science instrument operations during aerobraking were multifaceted. The GRS acquired a large amount of data, while THEMIS was used sparingly, and MARIE remained turned off after the anomaly occurred late in interplanetary cruise. In addition, significant atmospheric observations using the spacecraft accelerometers were collected during aerobraking.

Spacecraft geometries during aerobraking were not compatible with THEMIS operation. A spacecraft slew would have been required, and data downlink capability was extremely limited. However, a THEMIS atmospheric monitoring test

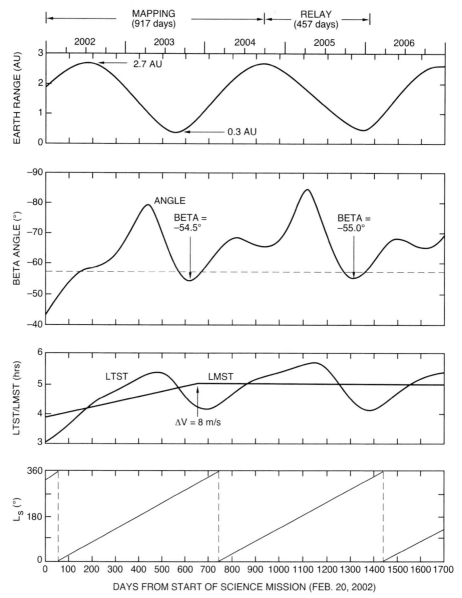

Figure 3. Expected Evolution of 2001 Mars Odyssey Mapping Orbit with Time. Earth-Mars range will vary between a minimum of 0.3 AU and 2.7 AU, with high data downlink volumes at the Earth-Mars range minima, and low data downlink volumes at the Earth-Mars range maxima. Beta angle varies between −44 degrees at the start of mapping with two minima near −80 degrees. The Gamma Subsystem (GSS) acquires quality data when the Beta angle is less than −57.5 degrees. Local mean solar time (LMST) varies from near 4 p.m. at the start of mapping and drifts up to 5 p.m. at about 670 days into the mapping mission. Local true solar time (LTST) varies from 3:00 p.m. to 5:45 p.m. over the mission. L_S is the Mars-centered solar longitude; 0° represents the start of northern spring. Thus, Odyssey's mapping phase started in late northern winter.

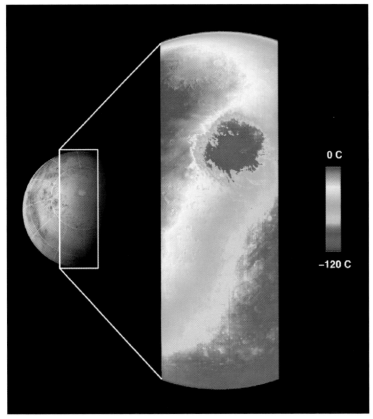

Figure 4. First THEMIS Infrared Image of Mars. This image of the south polar region was acquired during an atmospheric monitoring test just a few days after arrival using a single infrared filter centered at 10.1 μm. The prominent cold circular feature is the south polar carbon dioxide ice cap, about 300 km across. (JPL Public Release Photo PIA 03459)

was conducted following Mars orbit insertion (MOI) in order to demonstrate a capability for atmospheric monitoring in the event of an MGS TES failure during aerobraking (see Appendix B). The THEMIS atmospheric monitoring test on October 30, 2001 acquired THEMIS's first infrared image of Mars, at apoapsis (29,000 km) during the 9th orbit at Mars (Figure 4). This image was obtained when the spacecraft was looking toward the South Pole of Mars. The season on Mars was mid-summer in the Southern Hemisphere. The extremely cold (blue), circular feature in the center of the image is the Martian south polar carbon dioxide ice cap at a temperature of $-120°$C. The polar cap was more than 300 km in diameter at that time. The image spans the morning terminator covering a length of 6,500 km from limb to limb, with a resolution of approximately 7.7 km/pixel (4.6 mi/pixel). The cold region in the lower right portion of the image is associated with Argyre Basin.

Two of Odyssey's GRS instruments, the HEND and the NS, were turned on immediately after arrival at Mars and operated during most of the aerobraking phase. HEND operated continuously during aerobraking until the periapsis altitude was lowered to 180 km. Then, the HEND high voltage was cycled off/on at 20 minutes before/after periapsis. This protected against the possibility of arcing across the high voltage components in the electronics. HEND was also operated during the transition from aerobraking to mapping after the periapsis was raised above 180 kilometers. The NS was operated continuously and all NS channels detected the expected neutron signal from Mars until the periapsis altitude dropped to 180 km, at which time the NS was turned off. The NS was turned on again and operated during the transition from aerobraking to mapping, providing additional calibration for this instrument. HEND and NS observations immediately after arrival at Mars yielded indications that these two instruments were seeing neutrons from the surface. During the first two periapses at orbital altitudes of 300 km above the Martian North Pole, HEND detected large fluxes of neutrons from the surface of Mars. At the same time, the NS saw large variations in neutrons, varying with latitude for about three orbits at Mars after orbital insertion on October 24–25, 2001 (Feldman et al., 2002; Tokar et al., 2002).

A significant finding by the NS and HEND after aerobraking was the discovery that the flux of epithermal neutrons coming from the region of Mars poleward of 60° south latitude is depressed by at least a factor of 2 from that coming from the more equatorial latitudes. These results, shown in Figure 5, are consistent with near-surface polar terrain being rich in hydrogen within 30° of the pole (Feldman, 2002a). This new result for Mars can be placed in context by comparing it with the lunar epithermal flux measured by the NS on the Lunar Prospector spacecraft. Whereas the entire epithermal neutron flux from the Moon varies by only 15%, that measured along a single orbital track of Mars Odyssey varies by more than a factor of 2.

Another significant finding was provided by Odyssey's accelerometers, which measured atmospheric densities, scale heights, temperatures, and pressures over time as the spacecraft skimmed the upper portions of the Martian atmosphere during aerobraking. This provided the first in-situ evidence of winter polar warming in the Mars upper atmosphere (Keating et al., 2002 a,b). This warming may be due to a cross-equatorial meridional flow in the thermosphere from the summer hemisphere, which subsides in the winter polar region, bringing strong adiabatic heating. These data also showed winter polar temperatures rising with decreasing altitude, which agrees with some recent Mars atmospheric models. Also, temperatures at altitudes of 100 km near the winter pole were discovered to be twice as high as those predicted by earlier models.

In summary, the GRS, MARIE, and THEMIS operations in Earth-to-Mars planetary cruise and in the aerobraking transition phases before mapping demonstrated that these instruments were operating as planned and ready for routine opera-

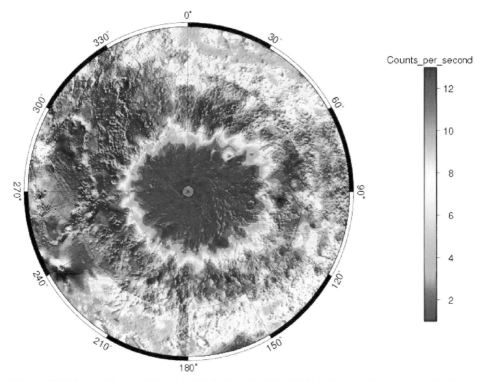

Figure 5. Epithermal Neutron Data for the Martian South Pole. Early Neutron Spectrometer (NS) and High-Energy Neutron Detector (HEND) observations indicate that the flux of epithermal neutrons coming from the region of Mars poleward of $\sim 60°$ south latitude is depressed from that coming from the more equatorial latitudes by at least a factor of two. These results are consistent with near surface polar terrain within 30 degrees of the pole being rich in hydrogen. See Feldman *et al.*, 2002a. (JPL Public Release Photo PIA 03487)

tions. In addition, atmospheric observations during aerobraking provided valuable insights into the dynamics of Mars upper atmosphere.

4.3. SCIENCE RESULTS–EARLY MAPPING

Odyssey observations in the first few months of mapping operation provided evidence for ice in the near-polar regions of Mars, discovered water ice at the surface near the south pole of Mars, and showed that radiation at Mars agrees with current models. GRS measurements showed that the uppermost meter of the Martian regolith near the poles (from latitudes of 60 degrees to the poles) was enriched in hydrogen. This was consistent with models that indicate this ice-rich layer underlies a hydrogen poor layer, and that water ice constitutes $35\% \pm 15\%$ of the ice-rich layer by mass. For regions near the South Pole, column densities of the hydrogen poor layer were about 150 grams per square centimeter at -42 latitude thinning to about 40 grams per square centimeter at -77 degrees south latitude

(Boynton *et al.*, July 2002, Feldman *et al.*, 2002b, and Mitrofanov *et al.*, 2002). Early THEMIS observations combined with earlier TES and Viking observations indicate that water ice is exposed near the edge of the perennial south polar cap of Mars (Titus *et al.*, 2003). Also, THEMIS visible images combined with images from the Mars Global Surveyor (MGS) suggest that melting snow may cause of the numerous eroded gullies first observed on Mars by the MGS's Mars Orbiting Camera (MOC) in 2000 (Christensen, 2003). Early MARIE observations indicate that the radiation dose rate at Mars is in good agreement with a model of the Galactic Cosmic Rays (Saganti *et al.*, 2003, this issue). Also, several Solar Particle Events (SPEs) were observed after March 2002 at Earth–Sun–Mars angles of 100° to 180°.

5. Odyssey Science Data Archiving and Distribution

A key requirement of the 2001 Mars Odyssey project is to provide data to the science community via the auspices of NASA's Planetary Data System (PDS). The instrument science teams are responsible for retrieving science packets (science telemetry), spacecraft planet instrument C-matrix events (SPICE) files, and other relevant information from the appropriate project databases, and transferring the files to their respective home institutions. Principal investigators (PIs) and their teams are responsible for generating engineering data records (EDRs) and reduced data records (RDRs) at each PI's home institution. Once EDR and RDR data products have been validated and released to the PDS as archives, the data and associated information will be made available to the research, education, and public communities.

Standard products form the core of the archives produced by Odyssey and released to the PDS for distribution to the science community and others. Standard products are well-defined, systematically generated data products, such as the EDRs and RDRs. These products and associated supporting information (e.g., documentation and index tables) will be validated and delivered to the PDS at regular intervals. The processes and schedules for generation and validation of standard products and archives, delivery to the PDS, and distribution to the science and other communities are described in greater detail in this section.

5.1. Generation, Validation and Delivery of Archives

The generation and validation of data products for the archives combines information from instrument science packets, spacecraft engineering packets, and other engineering information and data. In addition, SPICE kernels are generated and archived by the Navigation and Ancillary Information Facility (NAIF). Principal investigators access instrument science and engineering packets, and ancillary information such as spacecraft position and orientation in SPICE file format, from

TABLE IX
2001 Mars Odyssey EDR and Higher-Level Standard Data Products

Investigation	Product	Description
MARIE	MARIE-REDR	Raw times series of counts and radiation levels for MARIE detectors
	MARIE-RDR	Time series of radiation levels reduced to geophysical units
THEMIS	THM-VISEDR	Image cube of visible bands
	THM-IREDR	Image cube of infrared bands
	THM-VISRDR	Visible-band image cubes in radiance units
	THM-IRRDR	Infrared-band image cubes in radiance units
GRS	GRS-EDR	Raw gamma, NS, and HEND spectra
	GRS-IDR	Binned counts from GSR, NS, and HEND data
	GRS-RDR	Maps of element ratios and/or concentrations
SPICE	SPK	SPK (spacecraft) kernels
	PcK	PcK (planetary ephemeris) kernels
	IK	I (instruments) kernel
	CK	C (spacecraft rotations) kernels
	EK	E (experiment explanation/experimenter's notebook) kernels
Radio Science	ATDF	Archive Tracking Data Files
	ODF	Orbit Data Files
	RSR	Radio Science Receiver Records
Accelerometer	Altitude EDR	Constant Altitude Data

the project databases. All of these are used to generate NASA Level 1A experiment data records and higher-level derived products given in Table IX. These data products, along with supporting materials such as documentation, index tables, calibration files, algorithms, and/or software, form the Mars Odyssey science data archives. These archives are assembled under the auspices of the PIs and Instrument Team Leads with guidance and assistance (as needed) from the Mars Odyssey Interdisciplinary Scientist for data and archiving, and the Mars Odyssey Data and Products Working Group (DPWG). The DPWG helps by generating plans for the archives and by providing oversight during the archiving phase of the mission.

PDS policy requires that science archives be validated for scientific integrity and for compliance with PDS standards. This validation is done at several points along the path from receipt of raw packets to delivery of standard products using a combination of instrument team, mission, and PDS personnel. The instrument teams conduct primary validation of standard products and associated information as an integral part of their data analysis work. After validation, these standard products are assembled with supporting materials such as labels, index tables, documentation and software to form archives. Broad oversight of this validation work is done via the Science Data Validation Team (SDVT), a multi-mission team that ensures that all Mars Exploration Program projects are maintaining archiving quality and schedules. In addition, the Odyssey DPWG works on a detailed level to ensure that validation steps are accomplished.

An important step here is the validation of standard product archives before delivery to the PDS. Before the first archive delivery from an instrument, the standard products and supporting materials undergo a formal PDS peer review with participation of mission personnel, PDS personnel, as well as reviewers invited from the science community. The archives are examined for integrity of scientific content, compliance with the applicable data product Software Interface Specification (SIS) and archive SIS, and compliance with PDS standards. These peer reviews may result in requests for changes or additions to the supporting material in the archive ('liens'). The liens will be resolved before the archive can be accepted by PDS. Subsequent deliveries of archives throughout the mission are not required to undergo further peer review, as long as they do not vary substantially from the first delivery. They are, however, required to pass a validation check for PDS compliance. If minor errors are found, they may simply be documented in an errata file that accompanies the archive. Major errors will be corrected before the archive is accepted by PDS. After an instrument data archive has passed peer review and the PDS validation check, it is released ('delivered') to the PDS.

The PDS is also responsible for maintaining copies of its science archives on permanent physical media and for delivering copies of science archives to the National Space Science Data Center (NSSDC). As archives are released to the PDS, the receiving PDS Node will generate copies on appropriate physical media for long-term storage by PDS and NSSDC. During the six-month validation period before delivery and the interval following delivery during which the PDS Nodes are writing the archives to physical media, the data products will exist only as online archives. To reduce the risk of data loss, the Mars Odyssey Project is responsible for conducting periodic backups and maintaining redundant copies of online archives until they are permanently stored with PDS.

5.2. RELEASE AND DISTRIBUTION OF DATA PRODUCTS

The distribution of Odyssey data products is a two-step process. The 2001 Mars Odyssey project is responsible for making data products available to its own personnel and the instrument teams. The PDS is responsible for making data products available to the rest of the science community and the public. Whereas the data archives from previous missions have often been distributed to the science community on a set of physical media (e.g., CD ROMs), the large volume of data expected from Mars Odyssey makes this form of distribution expensive and impractical. Instead, distribution will be accomplished primarily by Internet access in ways that take advantage of the capabilities and expertise associated with PI home institution systems.

Because of differences in instrument data volumes and the facilities at the PI institutions, the distribution of Odyssey data products varies from instrument to instrument. In particular:

(a) The MARIE archives will be transferred to the PDS Planetary Plasma Interaction (PPI) Node at UCLA for online access once the archives have been validated and released.
(b) GRS archives will be transferred to the PDS Geosciences Node for online access. Custom-generated GRS products will be distributed online from the GRS facility at the University of Arizona, which will become a PDS Data Node for the duration of the mission and sometime beyond. When this Data Node is eventually dissolved, the custom product capability will be transferred to the PDS Geosciences Node.
(c) THEMIS archives will be distributed online from the THEMIS facility at Arizona State University, which will become a PDS Data Node for the duration of the mission and sometime beyond. When this Data Node is eventually dissolved, the THEMIS archives will be transferred to the PDS Imaging Node.

For all of these instruments, the PDS will generate hard copy volumes (primarily DVDs) as needed.

The 2001 Mars Odyssey Project releases integrated archives within six months of receipt of the last raw data included in the archives, in compliance with the Mars Exploration Program data release policy. During the six-month interval between receipt and release, the data are processed to standard products, validated through analyses, assembled into archives (online), and checked for compliance with PDS standards. The first Odyssey data release occurred in October 2002, six months after the first six weeks of the science mission and consisted of data acquired during those first six weeks of mapping. Thereafter, data releases will occur every three months. Each of the remaining deliveries will include three months worth of data acquired six to nine months previously.

In conclusion, these public PDS archives of Odyssey science data provide the general public and Mars science communities with a significant data set that fits into the current progression of Mars missions and fills important gaps in our knowledge of the planet.

Acknowledgements

This research was carried out at the Jet Propulsion Laboratory, California Institute of Technology, and was sponsored by the National Aeronautics and Space Administration.

Many have contributed to the success of the 2001 Mars Odyssey mission. The 2001 Mars Odyssey mission is managed by the Jet Propulsion Laboratory (JPL). Lockheed Martin Astronautics (LMA) built the Odyssey Orbiter in their Denver,

Colorado facilities. Both JPL and LMA have jointly operated the spacecraft from launch through interplanetary cruise and into orbit at Mars.

The Gamma Ray Spectrometer (GRS) instruments were integrated by the Lunar and Planetary Laboratory, University of Arizona, under the leadership of Prof. William Boynton, GRS Team Lead. The High-Energy Neutron Detector (HEND) was constructed by the Laboratory of Space Gamma Ray Spectroscopy at the Space Research Institute (Moscow, Russia) under the leadership of Igor Mitrofanov. The Neutron Spectrometer (NS) instrument was constructed by the Los Alamos National Laboratory (LANL) under the leadership of William Feldman. The Gamma Subsystem (GSS) was constructed by the Lunar and Planetary Laboratory, University of Arizona.

The Thermal Emission Imaging System (THEMIS) was supplied by the Arizona State University under the leadership of Prof. Philip Christensen. The Martian Radiation Environment Experiment (MARIE) instrument was supplied by the Johnson Space Center under the leadership of the late Gautam Badhwar.

The Planetary Data System (PDS) is acknowledged for guiding the Odyssey mission in developing plans for data archiving and for making Mars data available to the public and the science community.

Comments provided by reviewers contributed to a significant improvement of this report.

Appendix A. Odyssey Spacecraft–Details and Description

This appendix provides a detailed overview of the 2001 Mars Odyssey spacecraft. This orbiter was constructed to meet a number of requirements including providing power to the spacecraft systems via a solar array, providing a 6-m boom to separate the GRS from the spacecraft, providing communication to Earth via low-, medium-, and high-gain antennas, and providing a bus for all of the other science instruments, as well as normal spacecraft housekeeping components. The spacecraft bus, as shown in Figure 2, is a box measuring 2.2 m long, 1.7 m tall, and 2.6 m wide, composed mostly of aluminum and some titanium. At launch, Odyssey weighed 725.0 kg, including the 331.8-kg dry spacecraft with all of its subsystems, 348.7 kg of fuel and 44.2 kg of instruments.

A.1. COMMAND AND DATA HANDLING SUBSYSTEM

All of Odyssey's computing functions are performed by the command and data handling subsystem. The heart of this subsystem is a small onboard computer, a radiation-hardened version of the chips used on most personal computers. The subsystem runs Odyssey's flight software and controls the spacecraft through interface electronics with 128 MB of random access memory (RAM) and 3 MB of nonvolatile memory, which allows the system to maintain critical data even without power. Interface electronics make use of computer cards to communicate with external peripherals. These cards slip into slots in the computer's main board, giving the system specific functions it would not have otherwise. For redundancy purposes, there are two identical strings of these computer and interface electronics so that if one fails, the spacecraft can switch to the other.

There are a number of spacecraft interface cards with different functions. An interface card supports internal spacecraft communication with Odyssey's orientation sensors and its science instruments. A master input/output card collects signals from around the spacecraft and also sends commands to the electrical power subsystem. The uplink/downlink card provides an interface to Odyssey's telecommunications subsystems. There are two other boards in the command and data handling subsystem, both internally redundant. The module interface card controls when the spacecraft switches to backup hardware and serves as the spacecraft's time clock. A converter card takes electricity produced by the power subsystem and converts it into the proper voltages for the rest of the command and data handling subsystem components. The last interface card is a single, non-redundant, 1GB mass memory card that is used to store imaging data. The entire command and data handling subsystem weighs 11.1 kg.

A.2. Telecommunications

Odyssey's telecommunications subsystem is composed of a radio system which operates in the X-band microwave frequency range and in the ultra high frequency (UHF) range. This provides communication capability throughout all phases of the mission. The X-band system is used for communications between Earth and Odyssey, while the UHF system will be used for communications between the Odyssey orbiter and future Mars landers. The telecommunication subsystem weighs 23.9 kg.

A.3. Electrical Power

All of the spacecraft's power is generated, stored and distributed by the electrical power subsystem. The system obtains its power from an array of gallium arsenide solar cells on a panel measuring 7 m^2. A power distribution and drive unit contains switches that send power to various electrical loads around the spacecraft. Power is also stored in a 16-amp-hour nickel-hydrogen battery. The electrical power subsystem operates the gimbal drives on the high-gain antenna and the solar array. It also contains a pyro initiator unit, which fires pyrotechnically-actuated valves, activates burn wires, and opens and closes thruster valves. The electrical power subsystem weighs 86.0 kg.

A.4. Guidance, Navigation, and Control

The guidance, navigation, and control subsystem consists of three redundant pairs of sensors, which determine the spacecraft's orientation. A Sun sensor detects the position of the Sun as a backup to the star camera, while a star camera looks at star fields. Between star camera updates, the inertial measurement unit provides information on spacecraft orientation. This system also includes the reaction wheels, gyro-like devices used along with thrusters to control the spacecraft's orientation. There are a total of four reaction wheels, with three used for primary control and one as a backup. Odyssey's orientation is held fixed in relation to space ('three-axis stabilized') as opposed to being stabilized via spinning. The guidance, navigation, and control subsystem weighs 23.4 kg.

A.5. PROPULSION

The propulsion subsystem consists of a set of thrusters and a main engine. The thrusters are used to perform Odyssey's attitude control and trajectory correction maneuvers, while the main engine is used only once to place the spacecraft in orbit around Mars. The main engine, which uses hydrazine propellant with nitrogen tetroxide as an oxidizer, produces a minimum thrust of 695 Newtons. Each of the four thrusters used for attitude control produce a thrust of 0.9 Newtons. Four 22-Newton thrusters are used for turning the spacecraft. The propulsion subsystem also includes a single gaseous helium tank used to pressurize the fuel and oxidizer tanks, as well as miscellaneous tubing, pyro valves, and filters. The propulsion subsystem weighs 49.7 kg.

A.6. STRUCTURES

The spacecraft's structure consists of two modules: the propulsion module and the equipment module. The propulsion module contains tanks, thrusters, and associated plumbing. The equipment module is composed of an equipment deck and supports engineering components and the radiation experiment, and a science deck connected by struts. The topside of the science deck, in turn, supports the thermal emission imaging system, gamma ray spectrometer, the high-energy neutron detector, the neutron spectrometer and the star cameras. The underside of the science deck supports engineering components and the gamma ray spectrometer's central electronics box. The structure's subsystem weighs 81.7 kg.

A.7. THERMAL CONTROL AND MECHANISMS

The thermal control subsystem is responsible for maintaining the temperatures of the spacecraft components to within their allowable limits. This subsystem is a combination of heaters, radiators, louvers, blankets and thermal paint. The thermal control subsystem weighs 20.3 kg.

In addition to the thermal control subsystem, there are a number of mechanisms, several of which are associated with its high-gain antenna. Three retention and release devices are used to lock the antenna down during launch, cruise and aerobraking. Once the science orbit was attained, the antenna was released and deployed with a motor-driven hinge. The antenna's position is controlled with a two-axis gimbal assembly. There are also four retention and release devices used for the solar array. The three panels of the array are folded together and locked down for launch. After deployment, the solar array is also controlled using a two-axis gimbal assembly. The last mechanism is a retention and release device for the deployable 6-m boom for the Gamma Subsystem. All of these mechanisms combined weigh 24.2 kg.

A.8. Flight Software

Odyssey receives its commands via radio from Earth and translates them into spacecraft actions. To support this, the flight software is capable of running multiple concurrent sequences, as well as executing 'immediate' commands as soon as they are received. The software responsible for the data collection is extremely flexible. It collects data from the science and engineering devices and puts them in a number of holding bins, which can be modified via ground commands. The flight software is also responsible for a number of autonomous functions, such as attitude control and fault protection. If the software senses a fault, it will automatically perform a number of preset actions to put the spacecraft in a safe standby attitude, awaiting further direction from ground controllers.

A.9. Launch Vehicle

Odyssey was launched on a variant of Boeing's Delta II rocket, the 7925, which included nine strap-on solid-fuel motors. Each of the nine solid fuel boosters was 1 m in diameter and 13 m long; each contained 11,765 kg of fuel and provided a total thrust of 485,458 N at liftoff. The casings on these solid rocket motors were lightweight graphite epoxy. The first stage housed a main engine and two vernier engines. The vernier engines provided roll control during main engine burn and attitude control after main engine cutoff before the second stage separation. The main engine burned 96,000 kg of liquid fuel (a highly refined form of kerosene) and used liquid oxygen as an oxidizer.

The second stage was 2.4 m in diameter and 6 m long, which used 3,929 kg of liquid fuel, a 50/50 mixture of hydrazine and unsymmetric dimethly hydrazine. The oxidizer was 2,101 kg of nitrogen tetroxide. The second-stage engine performed two separate burns during the launch sequence.

The third and final stage of the Delta II 7925 provided the final thrust needed to place Odyssey on a trajectory to Mars. This upper stage was 1.25 m in diameter and consisted of a solid-fuel rocket motor with 2,012 kg of propellant and a nutation control system that provided stability after the motor ignited. A spin table attached to the top of the Delta's second stage supported and stabilized the Odyssey spacecraft and upper stage before it was separated from the second stage. The Odyssey spacecraft was mounted to the third stage by a payload attachment fitting. A yo-yo despin system decreased the spin rate of the spacecraft and upper stage before they

TABLE A1
2001 Mars Odyssey Spacecraft, Launch Vehicle, Mission and Web Site Information

Spacecraft	Dimensions: Main structure 2.2 m long, 1.7 m tall and 2.6 m wide; wingspan of solar array 5.7 m tip to tip
	Weight: 725 kg total, composed of 331.8 kg dry spacecraft, 348.7 kg of fuel of science instruments and 44.2 kg
	Science instruments: Thermal Emission Imaging System (THEMIS), Gamma Ray Spectrometer (GRS), Martian Radiation Environment Experiment (MARIE)
	Power: Solar array providing up to 1,500 W just after launch; 750 W at Mars
Launch Vehicle	Type: Delta II 7925
	Weight: 230,983 kg
Mission	Launch window: April 7–27, 2001 (launched on first opportunity – April 7, 2001)
	Earth-Mars distance at launch: 125 million km
	Total distance traveled Earth to Mars: 460 million km
	Mars arrival date: October 24, 2001
	Earth-Mars distance at arrival: 150 million km
	One-way speed of light time Mars-to-Earth at arrival: 8 minutes, 30 seconds
	Science mapping phase: February 2002 – August 2004
	Relay phase: August 2004 – November 2005
Web Sites	Information on 2001 Mars Odyssey Mission http://mars.jpl.nasa.gov/odyssey/ Information on THEMIS and latest released THEMIS Images http://themis.asu.edu/ Information on GRS suite and results http://grs.lpl.arizona.edu/ Information on MARIE and results http://marie.jsc.nasa.gov/ Odyssey Data Archives site http://wwwpds.wustl.edu/missions/odyssey/

separated from each other. During launch and ascent through Earth's atmosphere, the Odyssey spacecraft and upper stage was protected from aerodynamic forces by a 2.9-m-diameter payload fairing, which was jettisoned from the Delta II during second-stage powered flight at an altitude of 136 km.

A summary of information on the 2001 Mars Odyssey Spacecraft, Launch Vehicle, Mission and Web Sites is given in Table A-1.

Appendix B. Odyssey Spacecraft Operations: Launch, Interplanetary Cruise, and Aerobraking

This appendix provides details of Odyssey spacecraft operations from launch through interplanetary cruise and aerobraking.

B.1. Launch, Lift-Off, Insertion into a Parking Orbit

2001 Mars Odyssey was launched from Cape Canaveral on April 7, 2001 at 11:02 a.m. EDT during its very first launch opportunity. Odyssey lifted off from Space Launch Complex 17 at Cape Canaveral Air Station, Florida. Sixty-six seconds after launch, the first three solid rocket boosters were discarded followed by the next three boosters one second later. The final three boosters were jettisoned 2 minutes, 11 seconds after launch. About 4 minutes, 24 seconds after liftoff, the first stage stopped firing and was discarded 8 seconds later. About 5 seconds later the second stage engine ignited. The fairing or nose cone was discarded 4 minutes, 41 seconds after launch. The first burn of the second stage engine occurred at 10 minutes, 3 seconds after launch.

At this point the vehicle was in low Earth orbit at an altitude of 189 km and the second stage was restarted at 24 minutes, 32 seconds after launch. Small rockets were then fired to spin up the third stage on a turntable attached to the second stage. The third stage separated and ignited its motor, putting the spinning spacecraft on its interplanetary cruise trajectory. A nutation control system (a thruster on an arm mounted on the side of the third stage) was used to maintain stability during this third stage burn. After that, the spinning upper stage and the attached Odyssey spacecraft was despun so that the spacecraft could be separated and acquire its proper cruise orientation. This was accomplished by a set of weights that are reeled out from the spinning vehicle on flexible lines that act in a manner similar to spinning ice skaters slowing themselves by extending their arms. Odyssey separated from the Delta II third stage about 33 minutes after launch. Remaining spin was removed using the orbiter's onboard thrusters. At 36 minutes after launch, the solar array was unfolded and 8 minutes later it was locked into place. Then the spacecraft turned to its initial communication attitude and the radio transmitter was turned on. One hour after launch, the 34-m-diameter antenna at the Deep Space Network (DSN) complex near Canberra, Australia, acquired Odyssey's first signal.

B.2. Interplanetary Cruise

Odyssey's interplanetary cruise phase from Earth to Mars was accomplished in 200 days. Engineering activities during the cruise included checkout of the spacecraft in its cruise configuration, checkout and monitoring of the science instruments, and navigation activities necessary to determine and correct Odyssey's flight path to Mars. Science activities in this cruise phase included payload health and status checks, instrument calibrations, as well as data collection by the science instruments as spacecraft limitations allowed.

Odyssey's flight path from Earth to Mars was a Type 1 trajectory, taking less than 180 degrees around the Sun. During the first two months of interplanetary cruise, only the DSN station in Canberra was capable of viewing the spacecraft. Late in May, California's Goldstone station came into view, and by early June the Madrid station was also able to track the spacecraft. A small tracking station in Santiago, Chile was used during the first seven days following launch to fill in tracking coverage. During early Earth-to-Mars cruise, Odyssey transmitted to Earth using its medium-gain antenna and received commands on its low-gain antenna during the early portion on its flight. Later in cruise, Odyssey communicated via its high-gain antenna. Cruise command sequences were generated and uplinked every four weeks during regularly scheduled DSN passes.

The spacecraft determined its orientation in space during the interplanetary cruise phase via a star camera augmented with an inertial measurement unit. The spacecraft was oriented with its medium- or high-gain antenna pointed toward the Earth, while keeping the solar panels pointed toward the Sun. Spacecraft orientation was controlled by reaction wheels; devices similar to gyroscopes. These devices were occasionally 'desaturated', when their momentum was unloaded by firing the spacecraft's thrusters.

Navigation activities during interplanetary cruise involve the collection of two-way Doppler and ranging data as well as a series of delta differential one-way ranges (ΔDORs). The ΔDOR measurements were interferometric measurements between two radio sources: one of the radio sources being the differential one-way range tones from Odyssey and the second radio source was either a quasar (known, stable natural radio source) or the telemetry signal from the Mars Global Surveyor spacecraft. Both radio sources were recorded simultaneously at two earth-based radio antennas separated by about 120 degrees of longitude. Odyssey's position in the plane-of-the-sky was determined by triangulation of the ΔDOR signals, complementing the line-of-sight ranging provided by two-way Doppler and ranging data.

During interplanetary cruise, Odyssey conducted four trajectory course maneuvers (TCMs) by firing its thrusters to adjust its flight path. The first trajectory correction maneuver occurred 46 days after launch and corrected launch injection errors to adjust the Mars arrival aim point. It was followed by a second TCM 86 days after launch. The remaining two TCMs were 37 days and 13 days before arrival. The spacecraft was tracked by the DSN antennas for 24 hours before and after all of the TCMs. These maneuvers were conducted in a 'constrained turn-and-burn' mode in which the spacecraft turned to the desired burn attitude and fired the thrusters, while remaining in contact with Earth.

Science instruments were powered on, tested and calibrated during cruise. The THEMIS thermal imaging system took a picture of the Earth-Moon system 12 days after launch. Also, a THEMIS star calibration imaging was done 76 days after launch. There were two GRS calibration periods during which each of the suite's three sensors was operated. The MARIE was operated constantly during cruise until mid August, when it stopped responding to commands from Earth. This ended MARIE observations until after orbit insertion and aerobraking were completed.

Also, a test of the orbiter's UHF radio system was performed after launch. The 45-meter antenna at California's Stanford University was used to test the UHF system ability to receive and transmit. This UHF radio system will be used during Odyssey's relay phase to support future landers; it is not used as part of the orbiter's science mission. This test was successfully performed.

B.3. MARS ARRIVAL, ORBIT INSERTION, AND AEROBRAKING

Odyssey arrived at Mars on October 24, 2001. As it neared its closest point to the planet over the northern hemisphere, the spacecraft fired its 695-N main engine for 22 minutes allowing its capture in an elliptical orbit. Mars orbit insertion (MOI) performance was excellent. Navigation delivery was within 1 km of the target altitude. The post-MOI orbit period was 18.6 hours (the expected range was 15 –24 hours). Oxidizer burn-to-depletion was detected and triggered the burn cutoff as planned. The main engine thrust level was somewhat lower than expected (due to a better balanced spacecraft), resulting in a burn time of 1219 seconds versus an expected value of 1183 seconds. No period reduction maneuver following MOI was required.

Aerobraking provided a means of transitioning from the initial elliptical orbit immediately after arrival to the desired near circular mapping orbit. This technique was first demonstrated at Venus by Magellan and subsequently, at Mars by Mars Global Surveyor. It slows the spacecraft incrementally, orbit by orbit, by using frictional drag as it flies through the upper part of the planet's atmosphere. Friction from the atmosphere on the spacecraft and its wing-like solar array caused the spacecraft to lose some of its orbital energy during each periapsis, known as a 'drag pass'. As the spacecraft was slowed during each drag pass, the orbit was gradually lowered and circularized.

The Thermal Emission Spectrometer (TES) instrument on the Mars Global Surveyor spacecraft provided monitoring of the Mars atmosphere during Odyssey aerobraking. One important aspect of the TES observations was the potential to provide warning should a large dust storm have erupted. A dust storm could have increased the temperature of the atmosphere and resulted in a 'blooming' upward of the atmosphere. If this had occurred, the Odyssey spacecraft would have experienced a larger than expected drag force which could have overheated the spacecraft. With warning from TES, the Odyssey mission teams could adjust the height of orbit periapsis, raising Odyssey to a safe height. The THEMIS Mission Operations Team was prepared to turn on THEMIS and support atmospheric monitoring if the TES data were not available. Both the TES and THEMIS Atmospheric Science teams supported Odyssey's Atmospheric Advisory Group data interpretation. Fortunately, no significant dust storms occurred during aerobraking.

Aerobraking was conducted in three phases: walk-in, the main phase, and walk-out. The walk-in phase occurred during the first eight orbits following Mars arrival. The main aerobraking phase began once the point of the spacecraft's periapsis had been lowered to within 110 km above the Martian surface. As the spacecraft's orbit was reduced and circularized during 330 drag passes in 75 days, the periapsis moved northward, almost directly over Mars' North Pole. Small thruster firings when the spacecraft was at its apoapsis kept the drag pass altitude at the proper level to limit heating and dynamic pressure on the orbiter. The walk-out phase occurred during the last few days of aerobraking as described below. Aerobraking was completed with no major anomalies. Aerobraking was initiated 4 days after MOI and was completed after 75 days. Daily analyses of Mars atmosphere during aerobraking was done by the 2001 Mars Odyssey Atmospheric Advisory Group–led by Richard Zurek, JPL, assisted by Gerald Keating, George Washington University, and others at NASA's Langley Research Center in Hampton, Virginia.

The primary transition from aerobraking to a mapping orbit (the aerobraking walkout phase) was conducted in a series of three maneuvers in mid-January 2002. The first, aerobraking exit maneuver (ABX1) was conducted on January 11, 2002 to raise the orbit periapsis and terminate drag passes throughout the atmosphere. This ABX1 maneuver was performed at apoapsis and raised the periapsis altitude to about 200 km using a delta velocity (ΔV) of 20 m/s. The second aerobraking exit maneuver (ABX2), conducted on January 15, 2002, raised periapsis and changed the inclination. This was done when the argument of periapsis had drifted to the equator. This ABX2 maneuver raised periapsis altitude to 387 km and yielded an inclination that provided the proper local solar time drift for science mapping orbit. This maneuver was also conducted at apoapsis using a ΔV of 56 m/s. The third aerobraking exit maneuver (ABX3), conducted on January 17, 2002, reduced the orbit period and froze the orbit. This lowered apoapsis altitude to 450 km and rotated the argument of periapsis to the South Pole ($\omega = 270°$). This maneuver was conducted at apoapsis using a ΔV of 27 m/s.

Following these three aerobraking exit maneuvers, there were small orbit adjustments as well as the deployment of the high-gain antenna (HGA). Small, final orbit clean-up maneuvers on January 28 and 30, 2002 corrected residual biases and execution errors from the rocket burns associated with the three aerobraking exit maneuvers (ABX1, ABX2 and ABX3). These maneuvers adjusted both the periapsis and apoapsis altitudes using a ΔV of about 4 m/sec. After the final orbit was achieved, the last major spacecraft event before the start of mapping was the deployment of the HGA, which was successfully performed on February 4, 2002.

References

Badhwar, G. D.: 2004; Martian Radiation Environment Experiment (MARIE), *Space Science Reviews*, **110**, 131–142.
Boynton, W. V., Feldman, W. C., Squyres, S. W., Prettyman, T. H., Bruckner, J., Evans, L. G., Reedy, R. C., Starr, R., Arnold, J. R., Drake, D. M., Englert, P. A. J., Metzger, A. E., Mitrofanov, I., Trombka, J. I., d'Uston, C., Wänke, H., Gasnault, O., Hamara, D. K., Janes, D. M., Marcialis, R. L., Maurice, S., Mikheeva, I., Taylor, G. J., Tokar, R., and Shinohara, C.: 2002, 'Distribution of Hydrogen in the Near Surface of Mars: Evidence for Subsurface Ice Deposits', *Science* **297**, 81–85, July 5, 2002.
Boynton, W. V., Feldman, W. C., Mitrofanov, I., Evans, L. G., Reedy, R. C., Squyres, S. W., Starr, R., Trombka, J. I., d'Uston, C., Arnold, J. R., Englert, P. A. J., Metzger, A. E., Wänke, H., Bruckner, J., Drake, D. M., Shinohara, C., Hamara, D. K., and Fellows, C., 2004. 'The Mars Odyssey Gamma-Ray Spectrometer Instrument Suite', *Space Science Reviews*, **110**, 37–83.
Christensen, P. R., Jakosky, B. M., Kieffer, H. H., Malin, M. C., McSween, Jr., H. Y., Nealson, K. Mehall, G. L., Silverman, S. H., Ferry, S., and Caplinger, M.: 2004, 'The Thermal Emission Imaging System (THEMIS) for the Mars 2001 Odyssey Mission', *Space Science Reviews*, **110**, 85–130.
Christensen, P. R.: 2003; 'Formation of Recent Martian Gullies through Melting of Extensive Water-Ice Snow Deposits', *Nature* **422**, pp. 45–48, March 6, 2003.
Cucinotta, F. A.: 2004, 'In Memoriam, Gautam D. Badhwar (1940–2001)', *Space Science Reviews*, **110**, 157–159.
Cucinotta, F., Badhwar, G., Zeitlin, C., Cleghorn, T., Bahr, J., Beyer, T., Chambellan, C., Delaune, P., Dunn, R., Flanders, J., and Riman, F.: 2002, Exploration of the Mars Radiation Environment Using MARIE, 33rd Annual Lunar and Planetary Science Conference, March 11–15, 2002, Houston, Texas, abstract no. 1679.
Feldman, W. C., Prettyman, T. H., Tokar, R. L., Boynton, W. V., Byrd, R. C., Fuller, K. R., Gasnault, O., Longmire, J. L., Olsher, R. H., Storms, S. A., and Thornton, G. W.: 2001, The Fast Neutron Flux Spectrum Aboard Mars Odyssey During Cruise, American Geophysical Union, Fall Meeting 2001, abstract #P42A-0550.
Feldman, W. C., Tokar, R. L., Prettyman, T. H., Boynton, W. V., Moore, K. R., Gasnault, O., Lawson, S. L., Lawrence, D. J., and Elphic, R. C.: 2002a, Initial Results of the Mars Odyssey Neutron Spectrometer at Mars, 33rd Annual Lunar and Planetary Science Conference, March 11–15, 2002, Houston, Texas, abstract no. 1718.
Feldman, W. C., Boynton, W. V., Tokar, R. L., Prettyman, T. H., Gasnault, O., Squyres, S. W., Elphic, R. C., Lawrence, D. J., Lawson, S. L., Maurice, S., McKinney, G. W., Moore, K. R., and Reedy, R. C.: 2002b, 'Global Distribution of Neutrons from Mars: Results from Mars Odyssey', *Science* **297**, pp. 75–78, July 5, 2002.
Jakosky, B. M., Pelkey, S. M., Mellon, M. T., and Christensen, P. R.: 2001; Mars Surface Physical Properties: Mars Global Surveyor Results and Mars Odyssey Plans, American Geophysical Union, Fall Meeting 2001, abstract #P41A-04.
Keating, G. M., Tolson, R., Theriot, M., Hanna, J., Dwyer, A., Bougher, S., and Zurek, R.: 2002a, Detection of North Polar Winter Warming from the Mars Odyssey 2001 Accelerometer Experiment. 27th European Geophysical Society General Assembly, April 22–26, 2002, Nice, France.
Keating, G. M., Tolson, R., Theriot, M., Bougher, S., and Forget, F.: 2002b, Detection of North Polar Winter Warming from the Mars Odyssey 2001 Accelerometer Experiment. Proceedings of the 34th COSPAR Assembly/World Space Congress, C3.3, Planetary Atmosphere, Houston, Texas, October 2002.
Klug, S. L. and Christensen, P. R.: 2001, Involving Students in Active Planetary Research During the 2001 Mars Odyssey Mission: The THEMIS Student Imaging Project, 32nd Annual Lunar and Planetary Science Conference, March 12–16, 2001, Houston, Texas, abstract no. 1965.

Klug, S. L., Christensen, P. R., Watt, K., and Valderrama, P.: 2002, Using Planetary Data in Education: The Mars Global Surveyor and 2001 Mars Odyssey Missions, 33rd Annual Lunar and Planetary Science Conference, March 11–15, 2002, Houston, Texas, abstract no. 1758.

McCleese, D., Greeley, R., and MacPherson, G.: 2001, Science Planning for Exploring Mars. JPL Publication 01-7.

Mitrofanov, I., Anfimov, D., Kozyrev, A., Litvak, M., Sanin, A., Tret'yakov, V., Krylov, A., Shvetsov, V., Boynton, W., Shinohara, C., Hamara, D., and Saunders, R. S.: 2002, 'Maps of Subsurface Hydrogen from the High Energy Neutron Detector, Mars Odyssey', *Science* **297**, pp. 78–81, 5 July 2002.

Saganti, P. B., Cucinotta, F. A., Wilson, J. W., Simonsen, L. C., and Zeitlin, C.: 2004, 'Radiation Climate Map for Analyzing Risks to Astronauts on the Mars Surface from Galactic Cosmic Rays', *Space Science Reviews*, **110**, 143–156.

Saunders, R. S.: 2000, The Mars Surveyor Program - Planned Orbiter and Lander for 2001, 31st Annual Lunar and Planetary Science Conference, March 13–17, 2000, Houston, Texas, abstract no. 1776.

Saunders, R. S.: 2001a, 2001 Mars Odyssey Mission Science, American Geophysical Union, Fall Meeting 2001, abstract #P41A-08.

Saunders, R. S.: 2001b, Odyssey at Mars - Cruise and Aerobraking Science Summary, American Astronomical Society, DPS Meeting #33, abstract #48.07.

Saunders, R. S. and Meyer, M. A.: 2001, 2001 Mars Odyssey: Geologic Questions for Global Geochemical and Mineralogical Mapping, 32nd Annual Lunar and Planetary Science Conference, March 12–16, 2001, Houston, Texas, abstract no. 1945.

Saunders, R. S., Ahlf, P. R., Arvidson, R. E., Badhwar, G., Boynton, W. V., Christensen, P. R., Friedman, L. D., Kaplan, D., Malin, M., Meloy, T., Meyer, M., Mitrofonov, I. G., Smith, P., and Squyres, S. W.: 1999, Mars 2001 Mission: Science Overview, 30th Annual Lunar and Planetary Science Conference, March 15–29, 1999, Houston, TX, abstract no. 1769.

Titus, T. N., Kieffer, H. H., and Christensen, P. R.: 2003, 'Exposed Water Ice Discovered Near the South Pole of Mars', *Science*, **299**, 1048-1051, February 14, 2003.

Tokar, R. L., Feldman, W. C., Prettyman, T. P., Moore, K. R., Boynton, W. V., Gasnault, O., Lawson, S. L., Lawrence, D. J., and Elphic, R. C.: 2002, Comparison of Measured Thermal/Epithermal Neutron Flux and Simulation Predictions for the Odyssey Neutron Spectrometer in Orbit About Mars, 33rd Annual Lunar and Planetary Science Conference, March 11–15, 2002, Houston, Texas, abstract no. 1803.

Vuong, D., Badhwar, G., Cleghorn, T., and Wilson, T. L.: 2002, Anomalous Cosmic-Ray Candidates in MARIE Measurements, 33rd Annual Lunar and Planetary Science Conference, March 11–15, 2002, Houston, Texas, abstract no. 1652.

THE MARS ODYSSEY GAMMA-RAY SPECTROMETER INSTRUMENT SUITE

W. V. BOYNTON[1]*, W. C. FELDMAN[2], I. G. MITROFANOV[3], L. G. EVANS[4],
R. C. REEDY[5], S. W. SQUYRES[6], R. STARR[7], J. I. TROMBKA[8], C. D'USTON[9],
J. R. ARNOLD[10], P. A. J. ENGLERT[11], A. E. METZGER[12], H. WÄNKE[13],
J. BRÜCKNER[13], D. M. DRAKE[14], C. SHINOHARA[1], C. FELLOWS[1],
D. K. HAMARA[1], K. HARSHMAN[1], K. KERRY[1], C. TURNER[1], M. WARD[1],
H. BARTHE[9], K. R. FULLER[2], S. A. STORMS[2], G. W. THORNTON[2],
J. L. LONGMIRE[2], M. L. LITVAK[3] and A. K. TON'CHEV[3]

[1]*University of Arizona, Lunar and Planetary Laboratory, Tucson, AZ 85721, U.S.A.*
[2]*Los Alamos National Laboratory, Los Alamos, NM 87545, U.S.A.*
[3]*Space Research Institute, Moscow*
[4]*Science Programs, Computer Sciences Corporation, Lanham, Maryland 20706, U.S.A.*
[5]*Institute of Meteoritics, University of New Mexico, Albuquerque NM 87131, U.S.A.*
[6]*Cornell University, Center for Radiophysics & Space Research, Ithaca, NY 14853, U.S.A.*
[7]*Department of Physics, The Catholic University of America, Washington, DC 20064, U.S.A.*
[8]*NASA/Goddard Space Flight Center, Greenbelt, MD 20771, U.S.A.*
[9]*Centre d'Etude Spatiale des Rayonnements, Toulouse, France*
[10]*University of California San Diego, Department of Chemistry, La Jolla, CA 92093, U.S.A.*
[11]*University of Hawaii, Manoa, HI, U.S.A.*
[12]*Jet Propulsion Laboratory, California Institute of Technology, Pasadena, CA 91109, U.S.A.*
[13]*Max-Planck-Institut für Chemie, 6500 Mainz, Federal Republic of Germany*
[14]*TechSource, Sante Fe, NM 87505, U.S.A.*
(*Author for correspondence, E-mail: WBoynton@GAMMA1.lpl.arizona.edu)

(Received 13 September 2002; Accepted in final form 28 March 2003)

Abstract. The Mars Odyssey Gamma-Ray Spectrometer is a suite of three different instruments, a gamma subsystem (GS), a neutron spectrometer, and a high-energy neutron detector, working together to collect data that will permit the mapping of elemental concentrations on the surface of Mars. The instruments are complimentary in that the neutron instruments have greater sensitivity to low amounts of hydrogen, but their signals saturate as the hydrogen content gets high. The hydrogen signal in the GS, on the other hand, does not saturate at high hydrogen contents and is sensitive to small differences in hydrogen content even when hydrogen is very abundant. The hydrogen signal in the neutron instruments and the GS have a different dependence on depth, and thus by combining both data sets we can infer not only the amount of hydrogen, but constrain its distribution with depth. In addition to hydrogen, the GS determines the abundances of several other elements. The instruments, the basis of the technique, and the data processing requirements are described as are some expected applications of the data to scientific problems.

1. Introduction

The Mars Odyssey Gamma-Ray Spectrometer (GRS) is designed to record the spectra of gamma rays emitted from the martian surface as the spacecraft passes over different regions of the planet. The gamma rays arise from both radioactive decay and the nuclear interaction of elements with cosmic-ray particles. The energies of the gamma rays identify the elements responsible for the emissions, and their intensities determine the concentrations. In addition, the instrument also has the capability to determine the fluxes of thermal, epithermal, and high-energy neutrons coming from the surface of the planet. From these data, we can determine the ability of the surface materials to moderate and absorb neutrons and thus gain important information about the abundances and distributions of light elements, especially hydrogen. In addition, the neutron data are useful for determining the excitation flux for the gamma-rays made by nuclear reactions.

In this work, we first provide a simplified review of the basis of the gamma-ray and neutron spectrometry technique followed by a description of the GRS detectors and the analysis of their data. We then describe the instrumentation in detail and discuss its scientific objectives. Planetary gamma-ray and neutron spectroscopies were both proposed in the early 1960's by Arnold, Lingenfelter, and others (e.g., Lingenfelter *et al.*, 1961; Arnold *et al.*, 1962; Van Dilla *et al.*, 1962). On the Apollo 15 and 16 missions in 1971 and 1972, NaI(Tl) scintillation gamma-ray spectrometers were flown to the Moon, and spectra were accumulated over about 20% of the lunar surface. Abundances of magnesium, potassium, iron, titanium, and thorium were produced from the Apollo gamma-ray data (e.g., Bielefeld *et al.*, 1976; Davis 1980, Etchegaray-Ramirez *et al.*, 1983). The Apollo gamma-ray results confirmed that planetary gamma-ray spectroscopy is a useful tool in determining elemental compositions and that the theoretical calculations used in interpreting the data (e.g., Reedy, 1978) were basically correct. More recently, the Lunar Prospector mission flew both a bismuth germanate (BGO) gamma-ray instrument and a ^{3}He-based neutron instrument on a spacecraft in a low polar orbit (Feldman *et al.*, 1999). The Lunar Prospector gamma-ray spectrometer was used to make useful geochemical maps of the entire Moon (Lawrence *et al.*, 1999, 2002; Prettyman *et al.*, 2002), and its neutron instrument provided strong evidence for hydrogen near the lunar poles (Feldman *et al.*, 2001), probably in the form of ice trapped on the permanently-shadowed floors of polar impact craters.

Most recently, a gamma-ray spectrometer was flown to the asteroid Eros as part of the Near Earth Asteroid Rendezvous (NEAR) mission (Trombka *et al.*, 2000). This instrument was similar in many respects to the Apollo GRS instrument. However, instead of being mounted on a boom to reduce background from the spacecraft, the NEAR GRS was partially surrounded by a BGO detector that could be operated in anticoincidence mode to reject spacecraft background. The NEAR GRS produced only limited compositional results for Eros during the spacecraft's orbital mission, due to the low gamma-ray emission rate from the asteroid. However, the mission was terminated by gently landing the spacecraft on the asteroid. In the landed configuration the GRS was in direct physical contact with the asteroidal regolith, and the counting statistics improved markedly. The spacecraft survived in this configuration for a considerable period of time, and over 200 hours of gamma-ray data were collected, providing significant new information on the elemental composition of the asteroid at the landing location (Evans *et al.*, 2001).

Gamma-ray spectroscopy at Mars has been considered for many years (e.g., Metzger and Arnold, 1970), but has not yet been carried out with any success. The Soviet Mars-5 and Phobos missions included CsI(Tl) gamma-ray spectrometers that obtained limited data at Mars but few quantitative results on elemental abundances (Surkov, 1984; Surkov *et al.*, 1989; Trombka *et al.*, 1992). The Mars Observer spacecraft, launched in 1992, carried a combined gamma and neutron spectrometer (Boynton *et al.*, 1992) that was similar in many respects to the instrumentation carried on Mars Odyssey. However, Mars Observer was lost three days prior to orbit insertion, and no gamma-ray data from Mars were obtained. The Mars Odyssey GRS experiment is therefore the first highly capable elemental chemical mapping instrument to successfully orbit Mars.

Most papers on planetary gamma-ray spectroscopy consider only instruments that use inorganic scintillator technology: low-resolution spectrometers $\Delta E/E$. 8% full-width at half maximum at 662 keV) like NaI(Tl), CsI(Tl) and bismuth germanate (BGO). Substantial improvements in overall scientific performance are possible, however, via use of more advanced detector technology. The idea of using high-resolution solid-state gamma-ray spectrometers $\Delta E/E \approx 0.3\%$) made with detectors of germanium (Ge) dates back to Metzger *et al.* (1975), who discussed the virtues of having high spectral resolution even though the net efficiency of the detector for gamma rays is smaller than that of a scintillator spectrometer (cf. Metzger and Drake, 1990). Gamma-ray spectrometers using Ge detectors have been flown for astrophysical investigations for many years (e.g., Mahoney *et al.*, 1980). Several problems with the use of high-resolution gamma-ray spectrometers in space have been identified in simulation experiments, such as radiation damage (e.g., Pehl *et al.*, 1978; Brückner *et al.*, 1991) and unusual peaks in the spectra (e.g., Brückner *et al.*, 1987) and are discussed below.

2. Basis of the Technique

2.1. Sources and transport of neutrons

The ultimate source of almost all neutrons and most gamma rays in a planet are the galactic cosmic rays (GCR). The GCR are mainly (about 87%) protons (nuclei of ^{1}H), about 12% alpha particles (nuclei of ^{4}He), and $\sim 1\%$ heavier nuclei that arrive in the solar system with energies typically of ~ 0.1–10 GeV/nucleon. At these high energies, most of the GCR particles react before they are slowed much by energy-loss mechanisms when interacting with matter. The interaction of these GCR particles in a planet's surface results in a cascade of secondary particles, including ~ 10 neutrons, most with energies of $\sim 0.1-20$ MeV, per incident primary particle (Reedy and Arnold, 1972). Charged secondary particles, such as protons, with such energies are fairly rapidly stopped by ionization energy losses. Secondary neutrons, however, travel until they undergo an interaction with a nucleus in the planet's surface or escape to space.

Free neutrons (those outside a nucleus) are radioactive, beta-decaying to a proton and electron with a half-life of 615 seconds. Neutrons are made in planetary surfaces by reactions induced by cosmic-ray particles. The 'birth' of a neutron depends to some extent on the composition of the planet. More neutrons per incident cosmic-ray particle are produced from heavier elements, especially those with more neutrons than protons in their nuclei, such as titanium and iron (Drake et al., 1988; Masarik and Reedy, 1994). The flux of fast (about 0.6 to 8 MeV) neutrons measured by Lunar Prospector varied over the Moon's surface, with more fast neutrons being observed over regions with higher average atomic mass (Gasnault et al., 2001). The production of neutrons decreases roughly exponentially with depth in a planetary surface, with an e-folding length of ~ 150 g/cm^2.

The transport of neutrons in the planet depends very much on composition. The use of neutron spectroscopy to study the composition of a planet's surface was first discussed over four decades ago (Lingenfelter et al., 1961). Elastic scattering from nuclei slows neutrons, and the amount of slowing is dependent on the mass of the nucleus from which the neutron scatters; the lighter the nucleus, the more energy a neutron can lose per scatter (Fermi, 1950: Drake et al., 1988). The neutrons also have non-elastic interactions with nuclei. The cross sections for nuclei to interact with energetic (MeV) neutrons are all roughly the same, but the cross sections for the interaction with neutrons having thermal energies (below ~ 0.1 eV) or epithermal energies ($\sim 0.1-1000$ eV) can vary drastically among nuclei. For example, carbon and oxygen have very low absorption cross sections for neutrons with energies below ~ 1 keV, but iron, titanium, and chlorine all have very high cross sections for absorbing low-energy neutrons. The trace elements gadolinium and samarium have such huge cross sections that even at very low (a few parts per million by weight) elemental concentrations they can significantly affect neutron transport (e.g., Lingenfelter et al., 1972; Lapides, 1981).

A sizable fraction of neutrons made within the regolith are transported to the surface of a planet and escape to space. Most of these leakage neutrons are from the top few tens of centimeters (~ 100 g/cm^2) of the planet. Their energy spectrum reflects the transport properties in this top layer of the planet, including the martian atmosphere and any layers, such as water or carbon dioxide frost, deposited on the regolith. Leakage neutrons with energies below the gravitational binding energy (0.132 eV for Mars) can return to the surface (Feldman et al., 1989).

Neutrons with energies above the first excited levels of nuclei (~ 0.5 MeV for light nuclei) can interact with nuclei both by scattering elastically or by nonelastic scattering. Neutrons with energies above ~ 8 MeV can make more neutrons by reactions such as (n,2n), where two neutrons are emitted by a reaction induced by the first one. The flux of fast (~ 0.5–10 MeV) neutrons in Mars increases with depth near the very surface to a maximum near a depth of about 50 g/cm^2 and then decreases with depth (Masarik and Reedy, 1996).

2.1.1. *Neutron moderation*

The dominant energy-loss mechanism for neutrons having energies between about 0.1 eV and 0.5 MeV is elastic scattering. Energy loss, or moderation, results from the recoil of the struck nucleus leading to a constant, angle-averaged fractional energy reduction per collision. In the absence of other energy-loss mechanisms, such as absorption or leakage, continuous downscattering leads to an equilibrium neutron flux spectrum, which decreases with increasing energy as E^{-1}. This process, and the resultant energy spectrum, defines the epithermal component of the energy spectrum of neutrons.

Although the shape of the neutron flux spectrum in the epithermal range is determined by the elastic scattering process, its intensity is determined by the nuclear properties of the elements that constitute the planetary material. Most important are the atomic mass, which determines the amount of energy transferred from the neutron to the nucleus in each elastic collision, and the magnitude of the elastic cross section. Hydrogen is special in both respects; its mass is closely equal to that of the neutron, leading to an average transfer of half the neutron energy per collision, and the cross section for elastic n-p scattering is very high. Both effects promote fast moderation leading to a very low amplitude of the epithermal component of the equilibrium neutron spectrum. Detection of a significantly depressed epithermal amplitude therefore provides a unique signature of near-surface planetary hydrogen.

The dominant reaction types below about 0.1 eV are elastic collisions, which equilibrate the thermal motions of the neutron population with that of the nuclei of planetary matter, and neutron absorption reactions, which terminates the neutron. The resultant equilibrium neutron energy spectrum of these thermal neutrons is Maxwellian.

As with epithermal neutrons, the interaction processes for thermal neutrons define the Maxwellian shape of the equilibrium flux spectrum. However, the amplitude of the thermal-neutron flux is determined by the importance of absorption and leakage relative to the down-scattered injection rate. For these processes, carbon and oxygen are special. They both have extremely low neutron absorption cross sections yet moderately low atomic weights. Because thermal neutrons are gravitationally bound to Mars, the dominant neutron loss mechanism for a pure carbon dioxide deposit, such as is expected over the winter polar caps, is due to trace abundances of argon and nitrogen in the atmosphere and to neutron beta decay. Resultant thermal amplitudes should therefore be greatly enhanced (Drake *et al.*, 1988; Feldman *et al.*, 1993a). In general, however, the thermal neutron amplitude does not provide a unique signature of enhanced carbon and oxygen in surface material, such as might result from a large carbonate deposit, because of the possible presence of other elements such as Fe, Ti, and Cl, which have large neutron absorption cross sections. Identification of such deposits therefore requires a combined analysis of gamma-ray and neutron data for a complete specification of the surface chemistry (Feldman and Jakosky, 1991). The flux-versus-depth profile of thermal neutrons depends strongly on the concentration of hydrogen in the surface. With no H in the surface, the peak flux of thermal neutrons is at a depth of about 150 g/cm^2. With increasing H contents, the depth of the thermal-neutron peak moves towards the surface (Lapides, 1981). In Mars, the thermal-neutron flux at the peak increases until the H content is about 0.5%. For higher H contents, H is the dominant absorber of thermal neutrons, and the flux of thermal neutrons at the peak depth starts to decrease with increasing H content (Feldman et al., 1993a; Masarik and Reedy, 1996).

The neutron components of the Mars Odyssey Gamma-Ray Spectrometer will be used to determine the fluxes of thermal, epithermal, and fast neutrons for each martian surface resolution element. Numerical simulations have shown (e.g., Drake et al., 1988; Dagge et al., 1991; Feldman et al., 1993a) that the epithermal amplitude by itself is a sensitive indicator of the surface hydrogen content. The fluxes of epithermal neutrons measured by Lunar Prospector were used to infer the presence of H at the poles of the Moon (Feldman et al., 2001). The fluxes of thermal and epithermal neutrons escaping from the Moon were also used to constrain chemical compositions of lunar surface regions (Feldman et al., 2000). Vertical stratigraphy is also possible using a combination of neutron and gamma-ray data (e.g., Evans and Squyres, 1987, Boynton et al., 2002).

The numerical simulations mentioned above have also shown that the thermal neutron amplitude is a sensitive indicator of carbon dioxide frost. Although this result is not unique because of the effects of normally abundant neutron absorbing nuclei in the Martian regolith as noted above, the fact that the underlying surface composition will not change during the winter season should allow use of the neutron data alone to map the growth and decay of the carbon-dioxide frost polar caps.

2.2. SOURCES OF GAMMA RAYS

A large variety of mechanisms produce gamma rays in a planet. Some, such as bremsstrahlung by energetic charged particles and the decay of cosmic-ray-produced neutral pions, produce a spectrum that is a featureless continuum. The gamma rays of interest in mapping elemental abundances, however, are those photons emitted by excited nuclei with very specific energies, typically in the range of 0.2–10 MeV. The excited levels in each nucleus occur at specific quantized energies, and the energy of a gamma ray made by a transition between nuclear levels can usually identify which nucleus produced it. Nuclei can be excited directly by cosmic-ray particles, by inelastic scattering and capture of neutrons, or in the decay of radioactive nuclei. Details of the production and transport of these gamma rays can be found in Reedy *et al.* (1973), Reedy (1978), Evans and Squyres (1987), and Evans *et al.*(1993).

Energetic particles emitted sporadically from the Sun can also result in gamma-ray production (Reedy *et al.*, 1973), but such solar-particle-produced gamma rays are of very limited use in planetary gamma-ray spectroscopy. The high intensity of solar particles penetrate the gamma-ray detector, yielding a high background signal that makes it difficult to detect prompt gamma rays from the surface of the planet. In addition, because the energy of the solar particles are much lower than that of galactic cosmic rays, most solar particles stop or react in the martian atmosphere and do not reach the surface.

2.2.1. *Decay of natural radioelements*

Several elements have isotopes with half-lives long enough that many nuclei have not decayed since they were formed by nucleosynthetic processes over 4.6 Ga ago. Normal potassium includes 0.012% of the isotope ^{40}K, which has a half-life of 1.25 Ga. When ^{40}K decays, 10.7% of the time the first excited level of ^{40}Ar at an energy of 1.461 MeV is produced. This excited level almost immediately decays to the ^{40}Ar ground state by emitting a 1.461 MeV gamma ray. Other naturally radioactive elements that will be mapped using decay gamma rays are 14-Ga ^{232}Th, 4.47-Ga ^{238}U, and 0.70-Ga ^{235}U. Several isotopes in the decay chains of ^{232}Th and ^{238}U actually emit the gamma rays used to map those elements. Thorium is usually mapped using the 2.615 MeV gamma ray emitted by its ^{208}Tl daughter. Other strong gamma rays made in the ^{232}Th decay chain include 0.911 MeV (^{228}Ac), 0.583 MeV (^{208}Tl), and 0.239 MeV (^{212}Pb). In the decay chain for ^{238}U, the strongest fluxes of gamma rays are at 1.764, 1.120, and 0.609 MeV (all from ^{214}Bi) and 0.352 MeV (^{214}Pb).

If a naturally radioactive element is uniformly distributed in the top ~ 100 g/cm^2 of a planet's surface, the fluxes of its decay gamma rays escaping the surface depend only on basic nuclear data, such as half-lives, gamma-ray yields, and the gamma-ray-transport property of overlying media (which is usually not very sensitive to composition) (cf., Reedy *et al.*, 1973). Non-uniformity in composition as a function of depth in the top layer from which gamma rays escape can be detected because gamma rays of different energies are attenuated by different amounts.

2.2.2. *Nonelastic-scattering-produced gamma rays*

The GCR-produced neutrons discussed above are the major source of gamma rays for most other elements. GCR particles can induce a wide range of reactions. An example of a reaction induced by GCR particles is the production of radioactive ^{24}Na from aluminum by a reaction in which a neutron enters a ^{27}Al nucleus and an alpha particle exits leaving a ^{24}Na nucleus. The short-hand notation for this reaction is ^{27}Al(n,α)^{24}Na. Although ^{24}Na emits a pair of gamma rays in its decay, these gamma rays are not used to map aluminum, as reactions with magnesium and silicon also produce large amounts of ^{24}Na and several prompt reactions also can produce the same gamma rays. Gamma rays made by the decay of cosmic-ray-produced radionuclides, such as ^{24}Na, are seldom used in planetary gamma-ray spectroscopy (Reedy et al., 1973). Most elements are mapped using neutron-inelastic-scattering or neutron-capture reactions, which produce excited states of nuclei that decay rapidly to their ground states by emitting characteristic gamma rays.

Most elements produce gamma rays by nonelastic scattering reactions where an excited level in a nucleus is populated by an energetic particle and then almost immediately (typically of the order of a picosecond) decays to a lower level. For example, the ^{28}Si nucleus has its first excited state at an energy of 1.77903 MeV, and any neutron having energy greater than 1.843 MeV (to conserve momentum, some of the neutron's incident energy goes into kinetic energy of the product nucleus) can excite ^{28}Si to that level. The decay of this level in ^{28}Si results in a gamma ray of 1.77897 MeV (with a very small amount of energy, 0.06 keV, going to the ^{28}Si nucleus for momentum conservation). This population of an excited level in the initial nucleus is called an inelastic-scattering reaction. The notation for this inelastic-scattering reaction is ^{28}Si(n,n(γ))^{28}Si. If a neutron incident on a ^{28}Si nucleus has enough energy (above about 3 MeV), other types of nonelastic-scattering reactions can occur (such as the ^{28}Si(n,α)^{25}Mg reaction). A neutron with energy above 11.7 MeV that reacts with ^{28}Si can lead to the emission of both a neutron and alpha particle that results in the population of the first excited level in ^{24}Mg, which then rapidly emits a 1.369 MeV gamma ray. This ^{28}Si(n,nα(γ))^{24}Mg reaction cannot be used to map silicon but is actually an unwanted interference with the mapping of magnesium via the 1.369 MeV gamma ray from ^{24}Mg(n,n(γ))^{24}Mg reactions.

Almost all elements can produce inelastic-scattering gamma rays from neutron interactions, the main exceptions being the very lightest elements like H and He that do not have gamma-ray-emitting excited levels. Some elements, such as those whose most abundant isotope is 'even-even' (i.e., even numbers of both protons and neutrons, like ^{56}Fe), have a strong, dominant gamma ray made by neutron inelastic-scattering reactions. However, a few even-even nuclei, like ^{40}Ca, and most other nuclei, like ^{27}Al, usually do not produce one dominant inelastic-scattering gamma ray but emit a number of gamma rays, all with relatively low probability. Inelastic-scattering gamma rays that are good for mapping various elements include 6.129 MeV for oxygen, 1.369 MeV for magnesium, 1.014 MeV for aluminum, 1.779 MeV for silicon, 3.737 MeV for calcium, and 0.847 and 1.238 MeV for iron.

A feature shown by a few of these nonelastic-scattering gamma rays is that they are produced with a spread of energies due to Doppler broadening. An example of a Doppler-broadened gamma ray is the 4.438 MeV gamma from ^{12}C, which is emitted from the excited ^{12}C nucleus while that nucleus is still recoiling from the reaction that made it (cf., Brückner *et al.*, 1987). Experimental simulations (e.g., Brückner *et al.*, 1992) show that martian-surface gamma rays emitted from excited levels with life-times greater than ~ 0.5 ps should not be broadened. Many gamma rays from aluminum (e.g., many of those above 2 MeV) and most gamma rays from oxygen (the main exceptions being the gamma rays at 6.129 and 3.853 MeV) are Doppler broadened. Most inelastic-scattering gamma rays produced in the martian atmosphere are Doppler broadened, due to the longer time required for the recoiling nucleus to slow down. This feature will help to distinguish them from the narrow gamma-ray lines from the same reactions in the martian surface (Reedy, 1988).

The calculation of the fluxes of nonelastic-scattering gamma rays require knowledge of cross sections for their production over a range of energies (cf., Reedy, 1978) that extend from threshold to many tens or even hundreds of MeV. The energy flux and spectra of GCR particles, and the neutrons that they produce, also need to be known (cf., Masarik and Reedy, 1996). The improved thermalization of the neutrons in a hydrogen-rich environment is accompanied by a slight decrease in the flux of neutrons with energies of a few MeV. For example, in a soil with 5% water, the flux of inelastic-scattering gamma rays will decrease about 20% relative to the case with no water (Lapides, 1981). Semi-empirical models (Reedy *et al.*, 1973) or neutron-transport codes (Lapides, 1981; Evans and Squyres, 1987; Dagge *et al.*, 1991; Masarik and Reedy, 1996) have been used to calculate the fluxes of inelastic-scattering gamma rays that escape from a planet. Some experiments (e.g., Brückner *et al.*, 1987; Metzger *et al.*, 1986b; Brückner *et al.*, 1992) have simulated the production of gamma rays.

2.2.3. *Neutron-capture-produced gamma rays*

Besides the decay of naturally radioactive elements and neutron inelastic-scattering reactions, the other major source of gamma rays used for elemental mapping is from neutron capture. Many elements have high cross sections for capture of neutrons at thermal energies. In most cases after the low-energy neutron is absorbed, one or more gamma rays are emitted. For example, titanium in the Moon was mapped by the ^{48}Ti(n,$(\gamma)^{49}$Ti reaction, using mainly the 6.418 and 6.760 MeV gamma rays and iron was mapped using the 7.631 and 7.646 MeV gamma rays from neutron-capture reactions with ^{56}Fe. The elements best mapped by the neutron-capture gamma rays have high cross sections for the capture of a thermal neutron and have one or more gamma rays emitted in high yields. In Mars, such elements include chlorine at 6.111 MeV and hydrogen at 2.223 MeV. Gadolinium, on the other hand, has a very high cross section for neutron capture, but no gamma ray is emitted with a high yield, making the detection of Gd by neutron-capture gamma rays difficult.

The fluxes of neutron-capture gamma rays are harder to predict, because many elements with high neutron-capture cross sections can affect the thermalization and transport of neutrons. For example, when only 0.1% hydrogen is present, the flux of thermal neutrons is considerably increased and the peak of the depth-versus-flux distribution is moved significantly closer to the surface than when hydrogen is absent (e.g., Drake et al., 1988; Feldman et al., 1989). Neutron-transport codes are used to calculate the distribution of low-energy neutrons in a surface with a given composition (or in layers with different compositions) (e.g., Lingenfelter et al., 1972; Lapides, 1981; Evans and Squyres, 1987; Drake et al., 1988; Dagge et al., 1991; Masarik and Reedy, 1996). Because these differences in the flux of thermal neutrons affect all elements analyzed via neutron capture equally, the ratio of elements is much less sensitive to composition than absolute abundances. These elemental ratios can be normalized to absolute abundances by comparison of the abundances of Fe and Si determined via gamma rays made by both thermal capture and inelastic scattering reactions. These elements produce high fluxes of gamma rays by both processes. In the case of the Mars Odyssey investigation, however, we will be able to calculate the neutron fluxes on the surface directly from the measurements of thermal and fast neutron fluxes at the spacecraft, which will help to recover the absolute abundances directly.

2.3. TRANSPORT OF GAMMA RAYS

In addition to the rate at which gamma rays are produced, their transport from their sources to the detector must be considered. Gamma rays can be scattered by interaction with other atoms, losing energy in the process. These scattered gamma rays add to the continuum at lower energies in the spectrum and decrease the signal-to-noise ratio for the discrete lines and are generally not useful for the quantitative analysis of Mars composition. In some special cases, however, this continuum can be used to study or confirm compositional variations among measured spectra (Thakur, 1997). The exponential attenuation coefficients for gamma rays range from about 0.125 cm^2/g at 0.2 MeV to 0.063 cm^2/g at 1.0 MeV to 0.024 cm^2/g at 10 MeV. Gamma rays escaping into space without undergoing an energy-loss interaction, therefore come from depths of several tens of g/cm^2.

For the gamma-ray detector on Mars Odyssey, which is roughly equally sensitive to gamma rays incident from all directions, the planet from horizon to horizon is the source of gamma rays detected. For a planet with no atmosphere, the spatial resolution of an orbiting gamma-ray spectrometer varies slightly with the type of reaction producing the gamma ray as different types of reactions produce gamma rays with different depth distributions (Reedy et al., 1973). The radius of the spatial-resolution circle below an orbiting gamma-ray spectrometer is approximately one half of its altitude above the planet's surface (Figure 1).

An atmosphere will attenuate and collimate gamma rays that escape from the planet's surface, especially the low-energy gamma rays. The Earth's atmosphere is too thick (1000 g/cm^2) for surface gamma rays to penetrate, but the ~ 15 g/cm^2 thick martian atmosphere transmits a significant fraction of surface gamma rays (cf., Masarik and Reedy, 1996). The transmission of gamma rays passing through the atmosphere normal to the martian surface is about 25% at 0.5 MeV and about

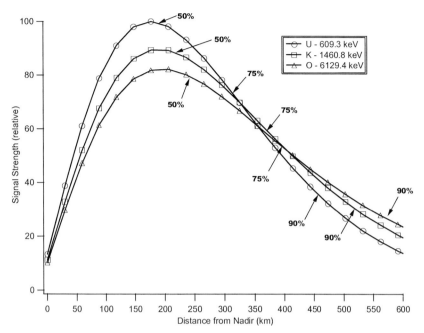

Figure 1. Signal intensity as a function of distance from the sub-spacecraft point (nadir) for three different gamma-ray energies. The signal increases with distance from nadir because the area of each annulus gets bigger, but it then drops off due to attenuation through greater path length through the regolith and atmosphere. The distance is indicated inside of which 50% of the signal is collected. This diagram is generated for a nominal Mars atmospheric thickness of 15 g/cm^2. At low elevations on Mars, where there is a thicker atmosphere, the spot size is smaller, and at higher elevations it is greater.

65% at 5 MeV. The average atmospheric transmission from all points on the Martian surface to the Mars Odyssey GRS at 400 km altitude, however, drops to about 7% at 0.5 MeV and 30% at 8 MeV because most of these gamma rays travel through more atmosphere. Since the gamma rays from near the horizon are most attenuated by the atmosphere, the spatial resolution for martian surface features is improved compared to the case of no atmosphere, and the amount of improvement is greatest for low energies and low orbits (Metzger and Arnold, 1970).

2.3.1. *Gamma-ray interferences*
There are some interferences with the gamma rays used to map elements in a planetary surface with other sources of the same gamma rays or gamma rays with energies similar enough that they will be difficult to separate in the analyses of gamma-ray spectra. For example, as noted above, the 1.369 MeV gamma ray from the ^{24}Mg(n,n(γ))^{24}Mg reaction will also be made by reactions with aluminum and silicon. The relative contributions from these other elements need to be known to get the component due only to magnesium. Another inelastic-scattering gamma ray that is produced in significant fluxes by other nonelastic-scattering reactions is the

4.438 MeV gamma ray from ^{12}C, which is readily made by the ^{16}O(n,nα(γ))^{12}C reaction (Reedy, 1978). Inelastic-scattering gamma rays for minor elements will often be overwhelmed by nonelastic-scattering gamma rays from major elements slightly higher in atomic mass and atomic number, such as the 1.434 MeV gamma ray in ^{52}Cr made from iron (cf., Reedy, 1978). For most minor elements, like chromium, neutron-capture reactions are best (such as its capture gamma ray at 8.885 MeV), as interferences are much less likely.

In a gamma-ray spectrometer with high energy resolution, such as that on the Mars Odyssey, the probabilities of spectral peak overlap are fairly rare, the 6.4196 MeV neutron-capture gamma ray from ^{40}Ca and the 6.4184 MeV one from ^{48}Ti being such a case. A more serious interference is from gamma rays produced in the gamma sensor head or the material surrounding it. For example, titanium and magnesium were used in the construction of the gamma sensor head and they will produce a significant flux of gamma rays that will significantly interfere with our ability to map these elements in the martian surface. These locally-produced gamma rays and ways to reduce or correct for such interferences from the instrument or spacecraft are discussed in Arnold *et al.* (1989). Backgrounds observed in the Mars Observer GRS during the cruise to Mars are presented in Boynton *et al.* (1998). Another source of background is from the energy deposited by charged cosmic-ray particles in the active region of the GRS (e.g., Evans *et al.*, 1998).

3. GRS Instrumentation

The GRS instrument suite consists of four components: the gamma sensor head (GSH), the neutron spectrometer sensor (NS), the high-energy neutron detector (HEND), and the central electronics box (CEB). The GSH is separated from the spacecraft by a 6-m boom, which was extended several months after the spacecraft entered the mapping orbit at Mars in order to minimize the spacecraft contribution to the gamma ray signal. The CEB houses the electronics for the gamma subsystem (GS), the NS, and various interface, power distribution, housekeeping and command and data-handling electronics. The HEND has its electronics contained in its sensor package but uses the CEB for command and control and as an electronic and data interface.

3.1. Gamma sensor head

A drawing of the GSH is shown in Figure 2. Its major components are the Ge detector assembly, the two-stage cooler subsystem, the door, and the Gamma Pulse Amplifier (GPA). It differs from the Mars Observer design in several important ways. First it uses a two-stage cooler versus the single-stage v-groove cooler of the Mars Observer instrument. Second the neutron subsystems are not an integral part of the sensor head as they were on Mars Observer. The neutron detector on

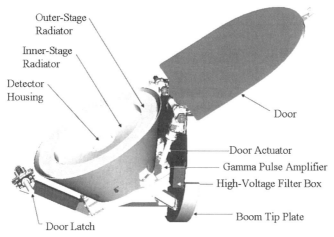
Figure 2. Drawing of the GRS gamma sensor head.

Mars Observer was incorporated as part of the anti-coincidence shield, which was designed to reduce the background due to charged particles. This charged-particle-rejection feature is not part of the Mars Odyssey GRS; it was removed as the result of a trade favoring greatly improved cooler performance. which yields greatly increased resistance to resolution degradation due to radiation damage (discussed below).

The solid-state detector is a large single crystal of n-type ultrahigh-purity germanium (HPGe), about 6.7 cm in diameter and 6.7 cm long, with semiconductor electrodes implanted or diffused such that the crystal becomes a diode, i.e. it will pass current in only one direction. The diode is operated in the reverse-bias mode with a potential of about 3000 V and a leakage current of less than 1 nA. The detector must be operated cold, less than about 140 K, to maintain high resolution and a low leakage current.

When a gamma ray interacts with the detector, hole-electron pairs are created that are quickly swept to the appropriate electrodes. This small charge is collected by a sensitive preamplifier, which produces a pulse whose height (voltage) is proportional to the energy deposited in the crystal. This pulse is then shaped and amplified and passed to a pulse-height analyzer, which counts the events in the form of a histogram sorted according to energy.

In the laboratory, HPGe detectors are generally operated near the boiling point of liquid nitrogen, 77 K; but in space it is difficult to attain this temperature, and the detector must be operated somewhat warmer. As mentioned above, detectors can normally operate at temperatures up to 140 K with little loss of energy resolution, but when they have been exposed to the radiation equivalent to about a one-year exposure in space, they need to be operated at 100 K or less (Brückner *et al.*, 1990). If they are irradiated with a sufficiently high dose of energetic particles, such as neutrons or protons, HPGe detectors will have their resolution degraded. During the mission the detector will be exposed to cosmic radiation for at least 5 years. As a result, the energy resolution of the detector will degrade to such a degree that it is substantially compromised for further measurements.

Brückner et al.(1991) incrementally exposed several large-volume n-type high-purity germanium detectors to a particle fluence of up to 10^8 protons/cm^2 (proton energy = 1.5 GeV) to induce radiation damage. The detectors were held at operating temperatures of 89, 100, and 120 K to cover temperature ranges expected for the mission. They found that the resolution degradation was correlated with higher operating temperature. In addition, the peak shapes in the recorded gamma ray spectra showed a significant change from a Gaussian shape to a broad complex structure. After a proton fluence equivalent to an exposure of 1 year in space, only the detector that was held at 89 K showed an energy resolution less than 3 keV; all other detectors had resolutions near or above 6 keV, a performance that is marginal for high-resolution spectroscopy (Brückner et al., 1990).

The radiation damage can be removed by heating the germanium crystal to temperatures of the order of 345 K for several days. The GS detector will operate at about 85 K, and thus, based on the above data, we expect to anneal it with on-board heaters about every two years based on normal GCR fluxes. Any increased fluence due to a coronal mass ejection (CME), however, could require us to anneal the detector more often.

3.2. Neutron Spectrometer Detector

The Mars Odyssey NS detector consists of a cubical block of boron-loaded plastic scintillator. It is segmented into four prism-shaped quadrants as shown in Figure 3. The prism segments are optically isolated from one another, and each is viewed by a separate 3.8-cm diameter photomultiplier tube (PMT). Both ends of the scintillator assembly are covered with a 0.069-cm thick sheet of cadmium to shield the ends of all prisms from thermal neutrons coming from those directions. In addition, the downward looking prism (N) has its downward looking face covered by an identical cadmium sheet so that it only responds to neutrons having energies larger than about 0.4 eV, which corresponds to the epithermal and fast neutron energy ranges.

Neutrons lose energy in the neutron detector through multiple elastic scattering collisions with the hydrogen and carbon nuclei that comprise the scintillator. Most of the energy is lost to proton recoils because protons and neutrons have close to the same mass and the cross section for (n,p) scattering is about four times larger than that for (n,^{12}C) at low energies. As the recoil protons slow down in the scintillator, they produce multiple ion-electron pairs that eventually recombine to produce photons. Collection of these photons by the PMTs produces pulses of charge that are then amplified and digitized by the neutron detector analog electronics to generate histograms. If the neutrons deposit all of their energy in the scintillator, they will eventually be captured by a ^{10}B nucleus to produce a second

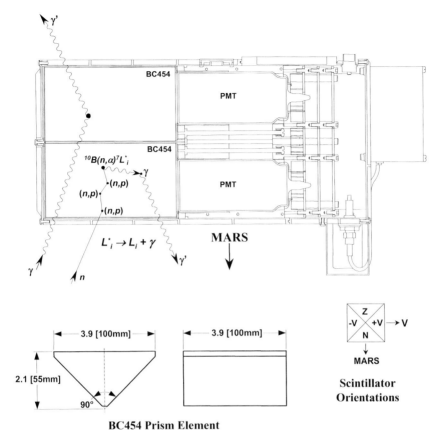

Figure 3. Drawing of the GRS neutron spectrometer sensor head showing two of the four BC454 prisms in cross section with their associated photomultiplier tubes (PMT). The four NS prisms are oriented with their faces viewing nadir (N), the direction of spacecraft velocity (+V), zenith (Z), and the direction opposite to spacecraft velocity (−V). A schematic of the interaction of neutrons and gamma rays with the boron-loaded plastic is also indicated.

pulse in the electronics. Although the Q-value of the ^{10}B(n,α)^{7}Li reaction is about 2.8 MeV; 478 keV of this energy goes to the gamma-ray de-excitation of the first excited state of ^{7}Li, which is populated 94% of the time in this reaction, and the remaining 2.3 MeV is split between the α and ^{7}Li recoils. Because of a pulse-height defect in the plastic scintillator, the recoil energy appears like a 93-keV electron. The sequence of events just described, is shown schematically in the upper part of Figure 3.

The signature of a thermal or epithermal neutron in the scintillator, is therefore a single pulse that has an amplitude that is characteristic of the ^{10}B(n,α)^{7}Li reaction. A histogram showing the response of one of the prisms to thermal neutrons measured before launch at Los Alamos National Laboratory, is shown in Figure 4. The single peak centered in channel 8 reflects the detection of the charged particle

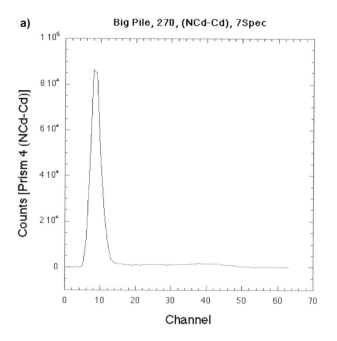

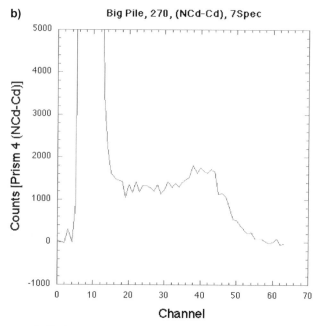

Figure 4. Histogram of NS photomultiplier tube output due to neutron interaction with one of the prisms. 4a. The peak is from the detection of the charged particle recoils from the $^{10}B(n,\alpha)^{7}Li$ reaction. 4b. The plateau at higher energies seen in this expanded scale is due to detection of both the charged particle recoil energy and energy lost by the 478 keV gamma ray from the ^{7}Li excited state.

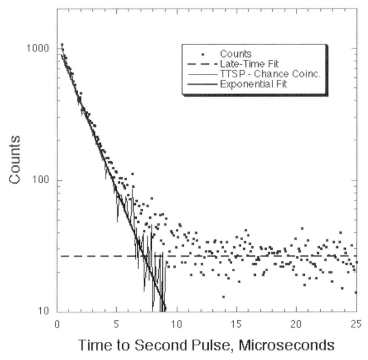

Figure 5. A time-to-second-pulse histogram using an AmB fast-neutron source. The counts show initially show an exponential decay followed by a constant count rate due to chance coincidences. The coincident rate is determined from the late counts and yield the horizontal dashed line. The raw counts are corrected for the chance coincidences by subtraction, yielding the points connected by the thin solid line. These corrected points are then fit by an exponential curve with an exponentiation time of ≈ 2 μs.

recoils alone (Figure 4a), and the broad plateau in the expanded-ordinate scale (Figure 4b) reflects the coincident detection of both the charged-particle recoil and the Compton interaction of the 478 keV gamma ray from ^{7}Li* in the same prism.

The signature of a fast neutron that has lost all of its energy in the scintillator is a time-correlated double pulse. The amplitude of the first pulse provides a measure of the energy of the neutron and that of the second pulse is the same as that for a thermal or epithermal neutron as shown in Figure 4. The time-to-second-pulse histogram for a calibration run using an AmB fast-neutron source at Los Alamos before launch is shown in Figure 5. The measured counts spectrum shows a constant counting rate at late times and a sharp upturn at early times, times less than about 8 μs after detection of a first interaction. The late-time portion of the spectrum reflects chance coincidences and can be fit by a constant, shown by the horizontal dashed line. A constant chance-coincident counting rate is expected for the low total counting rate that was present during this calibration run. Subtraction of the chance-coincidence background from the early-time counting rates yields an exponential decay (Figure 5). It has an exponentiation time of ~ 2 μs, which is

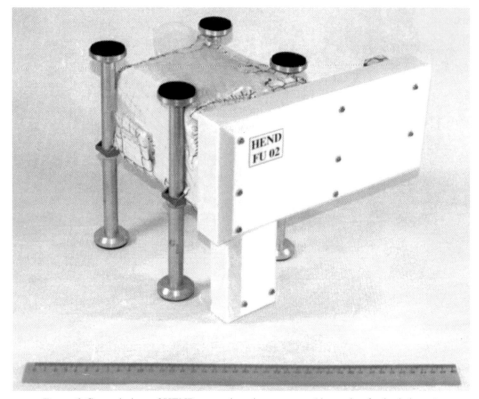

Figure 6. General view of HEND on engineering supports (the scale of ruler is in cm).

expected for a plastic scintillator containing 5% natural boron by weight as used in the Mars Odyssey neutron spectrometer.

The orientation of the outward normals to each of the four prism elements of the neutron spectrometer relative to the spacecraft velocity vector will be fixed throughout the mapping orbit. One face will view forward, one will view backward, one views downward and one views upward. These orientations are shown in the inset diagram to the lower right in Figure 3. Separation of the thermal and epithermal components will be possible using the relative counting rates of the forward and backward-directed prisms and a Doppler-filter technique (Feldman and Drake, 1986). Such a separation is made possible by the fact that the Mars Odyssey spacecraft will travel faster (3.4 km s^{-1}) than a thermal neutron (2.2 km s^{-1}) while in mapping orbit. The forward directed prism will therefore scoop up thermal neutrons and the backward directed one will outrun them. The difference in counting rates between forward- and backward-directed faces thus yields a measure of the flux of thermal neutrons. A measure of epithermal neutrons is provided by the downward facing prism because it is completely shielded from the outside by sheets of cadmium and the other three prisms.

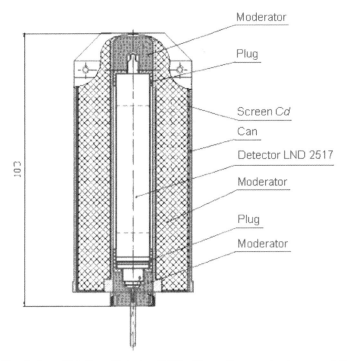

Figure 7. a) The design of Medium Detector (MD) with medium moderator around the LND 2517 ^{3}He proportional counter. b) The design of the Large Detector (LD) with a thick moderator around the LND 2517 ^{3}He proportional counter. c) The design of the scintillation block showing the internal scintillator, SC/IN with crystal of stylbene and PMT R1924, and the external scintillator, SC/OUT with crystal of CsI and PMT R1840.

3.3. HIGH ENERGY NEUTRON DETECTOR

The High-Energy Neutron Detector (HEND) was developed in the laboratory of Space Gamma-Ray Spectroscopy of the Space Research Institute (Moscow, Russia). HEND (Figure 6) integrates in one unit a set of five different sensors and electronics boards. The set of sensors includes three detectors with ^{3}He proportional counters and a scintillation block with two scintillators.

The detectors with proportional counters SD (Small Detector), MD (Medium Detector) and LD (Large Detector), are based on the industrial ^{3}He counter LND2517 surrounded by thin, medium and thick moderators of polyethylene inside cadmium cans, respectively (Figure 7). This counter is most efficient for thermal neutrons, but the cadmium can shields all external thermal and epi-thermal neutrons with energies below 0.4 eV. The counters detects neutrons with higher energies that are moderated from higher energies in the surrounding polyethylene. These detectors thus have different efficiencies for neutrons at different energy ranges depending on the thickness of moderators.

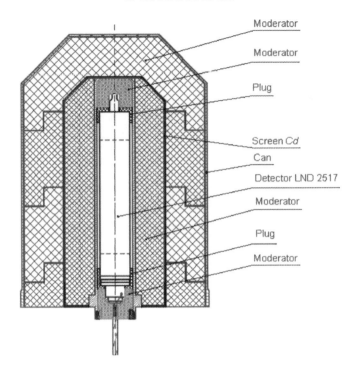

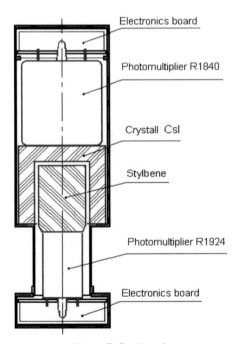

Figure 7. Continued.

Detector SD is most sensitive for neutrons at the energy range 0.4 eV – 1 keV; MD is the most sensitive at the energy range 0.4 eV – 100 keV; LD is most sensitive for neutrons at the energy range 10 eV–1 MeV. These three detectors SD, MD and LD, produce three separate signals SSD, SMD and SLD, respectively.

The scintillation block (Figure 7) contains an internal scintillator SC/IN with a stylbene crystal and a Hamamatsu photo multiplier tube (PMT) R1924. High energy neutrons produce recoil protons in stylbene, which are detected as proton-based counts at the energy range 800 keV – 15 MeV. The stylbene is also sensitive to gamma-ray photons, which are detected as electron-based counts at the energy range 60 keV – 2 MeV. HEND has the analog electronics of SC/IN detector, which separates proton-based counts from electron-based counts. This detector produces two separate signals $S_{SC/IN/N}$ and $S_{SC/IN/G}$ for neutrons and gamma-ray photons, respectively. The separation is based on the measurable difference between time profiles of scintillation light from protons and electrons. The efficiency of separation was directly measured under conditions when the SC/IN detector measured strong radioactive sources of gamma-rays in the MeV energy range. A number of neutron-like counts detected from this source resulted from the false identification of electron-based signals as proton-based signals. The efficiency of this separation corresponds to one false separation of a proton-based count from 2000 electron-based counts.

The scintillation lock (Figure 7) also contains an external scintillator, SC/OUT, with a CsI crystal and a Hamamatsu PMT-R1840 for detection of charge particles and gamma-rays above 30 keV. This detector provides the analog signal SSC/OUT and the digital veto signal for anti-coincidence rejection of protons, which could be detected in SC/IN as proton-based counts.

All six analog signals S_{SD}, S_{MD}, S_{LD}, $S_{SC/IN/N}$, $S_{SC/IN/G}$ and $S_{SC/OUT}$ are digitized into 16 energy channels. The energy spectra of these signals can be accumulated over time intervals from 12 sec up to 1 hr, but normally the start and end of the accumulation is synchronized by the GRS CEB to the same time period as the spectra of the GS.

In the case of a cosmic gamma-ray burst or solar flare, HEND has an automatic triggering unit that switches on the burst-detection mode when count rates of signals $S_{SC/IN/G}$ and/or $S_{SC/OUT}$ exceed some pre-selected thresholds. These thresholds are implemented by commands and require that the number of $S_{SC/IN/G}$ counts during 0.25 sec and/or number of $S_{SC/OUT}$ counts during 1.0 sec are significantly higher than the levels of background fluctuations. HEND continuously records 60-sec-long time profiles of signals $S_{SC/IN/G}$ and $S_{SC/OUT}$ with a time resolution 0.25 sec and 1.0 sec, respectively. If triggering took place during the accumulation of a current profile, these data are output at the next synchronization pulse from the CEB.

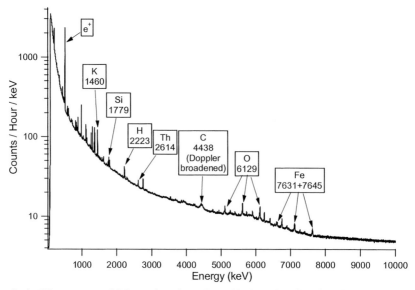

Figure 8. A GS spectrum of Mars taken from June 10 through July 16, 2002. Several emission lines are labeled with their energy in keV and the element responsible for the line. This shows our full-scale energy range of 10 MeV at our nominal gain setting. The continuum above about 8 MeV is due mostly to charged particle interactions in the detector. The broad continuum at lower energies is due mostly to scattered gamma rays that have lost a fraction of their energy. Scattering can occur in the regolith, the atmosphere, or the instrument itself. The line labeled e^+ is due to one of the two 511-keV gamma rays that occur when positrons and electrons annihilate. Positrons are made in one of the processes by which high-energy gamma-rays can interact with matter. The high-energy lines in the spectrum occur in threes, with the lines separated by 511 keV. The lower-energy lines are due to the loss of one or both of the 511 keV gamma rays made when a high-energy photon interacts with the detector via the pair production process.

3.4. THE CENTRAL ELECTRONICS BOX

The CEB contains all of the electronics associated with the GRS, except for the preamplifiers and the HEND electronics. It contains the power supplies, the main amplifiers, the analog-to-digital converters (ADCs), the GS and NS pulse-height analyzers (PHAs), and the central processing units and memory. It is mounted inside the spacecraft on the bottom of the instrument deck. The GS and NS PHAs generate a spectrum by counting each photon that the detector sees. They contain ADCs that digitize the height of each pulse, generating a number that is proportional to the energy deposited in the detector. This number is then used as an address of a memory location, or channel, corresponding to that energy. The PHAs increment the value stored in that location and thus generate spectra that are histograms of the number of events in each energy channel. The GS, which has a large energy range and high spectral resolution, uses 16,384 channels to store its spectrum, and the NS uses only 16 channels. A sample spectrum taken shortly after arriving at Mars is shown in Figure 8.

TABLE I
Significant GRS events in mission timeline.

Operation	Begin	End
GRS power on	25-Apr-2001	
NS and HEND data collection	2-May-2001	
GS door open, GS data collection	27-Jun-2001	
GS off due to coronal mass ejection	16-Aug-2001	17-Aug-2001
GS door close for MOI	31-Aug-2001	
GRS power off for MOI	24-Sep-2001	
GRS post-aerobreaking chek out	14-Jan-2002	
Spacecraft in mapping configuration	19-Feb-2002	
GS door open, GS data collection	20-Feb-2002	
GS warm anneal at 323 K	8-Mar-2002	22-Mar-2002
GS warmer anneal at 345 K	6-May-2002	21-May-2002
GS door close for boom erection	1-Jun-2002	
GS door open, GS data collection	6-Jun-2002	

The GS also has several counters that are used to monitor count rates in several energy bands. One is defined by the lower-level discriminator (LLD), which is setable in the range from 0 to 1000 keV and is used to determine which events are to be digitized by the ADC. There is also an upper-level discriminator (ULD), which has an energy range from 1 to 24 MeV and determines the maximum energy that the ADC will digitize. By setting the LLD and ULD to appropriate values, the ADC will not waste time digitizing low-energy noise or high-energy gamma-rays that would be off scale. Three other discriminators define energy windows, E2-E1, E3-E2, and >E3, that are used to define energy bands for triggering on gamma-ray bursts. When the count rate in any of these energy windows exceed a setable number of standard deviations above the mean value, the GS will switch into burst mode and return count-rate data in these windows with a time resolution setable from 10 to 64 msec.

The CEB also collects a variety of engineering data from the NS and GS. These include temperatures, voltages, currents and a variety of status bits. The temperature data of some of the GS subsystems are very important, as they are used to correct the data for temperature drift as discussed in the next section.

4. Mission Operations and Data Analysis

4.1. MISSION OPERATIONS

The Mars Odyssey Mission is described elsewhere (Saunders *et al.*, 2004). Just the events as they relate to the GRS will be discussed here. Table I provides a time line of significant events during cruise and early mapping. At 18 days into cruise, the GRS was powered on (CEB only), and the three instrument subsystems were powered on and data collection was begun 7 days later. At this time the gamma detector was still warm, so no gamma data was returned. On day 79 the door on the GSH was opened allowing the sensor to cool. After two days the detector had cooled sufficiently that the high voltage could be applied and GS data collection began. The instrument was turned off twice during cruise for spacecraft operation testing. At one point late in cruise the GS shut down autonomously, as intended, due to a high-current condition caused by a coronal mass-ejection (CME) event. The instrument is designed to remove the high-voltage bias on the detector if the ground-selectable setpoints on the detector temperature or current are exceeded. The bias was reapplied two days later and data collection resumed. At 55 days before arrival at Mars, the high voltage to the GSH was removed and its door was closed, so the spacecraft team did not have to be concerned about its closure in the final stages of preparation for Mars orbit insertion (MOI). Finally, at 30 days before arrival, the entire instrument suite was powered down until after MOI.

The mapping phase began on Feb 19, 2002. We turned on the GRS and began to collect data from HEND and NS during instrument check out 36 days earlier. We opened the GS door one day earlier and began to collect gamma data until one day after mapping had begun. We found the resolution of the GS detector had suffered badly due to a very large CME event that occurred after we shut down in preparation for MOI. We annealed the detector twice, first at a modest temperature of 323 K, and later, in preparation for boom deployment, we did it again at our maximum qualification temperature of 345 K. The first anneal was moderately effective at removing the radiation damage, but we still had significant degradation that we were able to remove with the higher-temperature anneal. The second anneal improved the resolution at 1461 keV from 5.0 keV (full width at half maxium) and 15.1 keV (full width at tenth maximum) to 3.9 and 11.1, respectively.

4.2. REDUCTION OF GS DATA

The purpose of the cruise data collection is both to check out instrument performance and to measure the spacecraft background. In addition, some cruise science can be performed with the detection of cosmic gamma-ray bursts and solar flares. The materials in the spacecraft can have a significant effect on the signal in both the neutron and the gamma sensors. The spacecraft provides mass with which the cosmic rays can interact to produce fast neutrons via the same spallation process that happens on Mars. There is only a modest amount of mass however, so many neutrons escape without being moderated. There is a large amount of fuel in the spacecraft during cruise, and the hydrogen in it is very effective at moderating neutrons. The fuel, N_2H_2, also has a large amount of nitrogen, which is a very effective absorber of neutrons. Consequently during cruise the thermal neutron flux measured near the spacecraft was very low (Feldman et al., 2002b).

The elements in the spacecraft will also give off gamma rays via the same process described above for Mars itself. By being close to the spacecraft we can determine the signal from the spacecraft with greater sensitivity than its contribution in Mars orbit. On the other hand, because the thermal neutron excitation flux during cruise was so low, even with the greater sensitivity we could not detect any thermal neutron capture gamma rays except for a very weak one due to hydrogen. For the gamma lines formed via fast-neutron processes, the cruise data will be a good measure of the spacecraft background.

A thorough discussion of background subtraction is beyond the scope of this work. We make use of the extendable boom to measure the signal in orbit both before and after the boom extension, but this extension only partially helps us to make a background correction. The background is complicated by the fact that we also have a background contribution from the instrument construction materials, and obviously when we extend the boom, our sensor does not move away from this source of background. In addition, some elements provide such a small signal that we may not be able to detect them during the period before boom extension. Another means of determining the background that appears to work well for strong lines, is based on the signal over the winter poles. The polar regions in winter are covered by a thick CO_2 frost that is opaque to gamma rays, so except for lines due to carbon and oxygen, we get a good measure of our background, including lines from the instrument, when we are over these regions.

Figure 8 shows a GS spectrum taken after boom deployment. It shows the characteristic sharp gamma-ray emission lines superimposed on a continuum. The lines reflect gamma rays from discrete nuclear transitions that pass unscattered to the detector through the intervening material. This material includes the regolith, the atmosphere and the materials used in the GSH itself. Most of the gamma rays will be Compton scattered on their way to the detector, which causes them to lose a fraction of their energy and thus they do not get counted as an event in the full-energy peak of the spectrum. Much of the continuum in the spectrum is produced by these scattered gamma rays. Owing to the finite energy resolution of the detector, the lines are recorded as peaks that are generally Gaussian in shape. The spectra are analyzed by determining the energy and the area of each peak of interest. The energy identifies the element responsible for the gamma ray emission, and the area above the continuum (number of counts) is proportional to the amount of that element in the surface. The determination of the proportionality constant that relates these counts in the spectrum to the concentration of the element is far from trivial, however.

The precise determination of the peak area in a gamma ray spectrum is important for the determination of element concentrations. There are a number of peak-fitting routines, which perform a peak-area determination for laboratory gamma ray spectra. In prompt gamma ray spectrometry the spectral shape of the gamma-ray spectra is more complex, and certain peaks differ from the normal narrow Gaussian shape. Doppler broadening, caused by a rapid emission of a gamma ray from a fast-moving excited nucleus, is usually the reason for these oddly shaped peaks. In addition, build-up of radiation damage in the HPGe detector can give rise to an asymmetrical peak broadening.

The number of counts in a peak depends on the efficiency of the detector. The efficiency is the probability that a photon of a certain energy striking the detector will register a count in the peak corresponding to that photon energy. Typical values range from about 20% at 1 MeV to 1.5% at 10 MeV. The losses are due to gamma rays passing through the detector with no interaction or scattering inside the detector with only a partial energy deposition. The flux of photons entering the sensor then must be corrected for the attenuation from the intervening material, which includes the detector packaging hardware as well as the planetary atmosphere, and for the solid angle of the planet relative to the instrument.

By far the most difficult task is the calculation of the relationship between the concentration of the element and the gamma-ray flux at the surface from that element. This relationship depends on the succession of processes that are not measured directly but must be modeled or inferred from the measured gamma-ray spectra. Because the neutron flux is dependent on composition, models are first used to estimate the concentration, and model calculations are then made to determine the neutron and gamma spatial and energy distributions from which a revised composition can be calculated. An iterative process is then used to determine both the final elemental composition and the neutron flux.

4.3. SPATIAL RESOLUTION AND SENSITIVITIES

A key characteristic of the GRS data will be that individual spectra, typically acquired during $\sim$ 20-s integration intervals, will have inadequate signal-to-noise ratio to be statistically significant. Instead, spectra taken at different times over the same region must be summed before spectral line intensities can be derived. It takes many hours of accumulated data to yield a spectrum that can be analyzed to return compositional results with reasonable uncertainty. As noted earlier, the diameter of a spatial resolution element (from which 50% of the signal is received) on the Martian surface is about 360 to 450 km depending on the gamma-ray energy (Figure 1). Table II shows the time available to measure gamma rays above one spatial resolution element for the complete 917-day Mars Odyssey mission as a function of latitude. Due to the higher density of orbital trajectories at the poles, the time spent over a given resolution element at high latitudes ($>80°$) is about 50

TABLE II

Accumulation times in a 450-km GRS footprint over the course of the 917-day Odyssey Mission.

Latitude (deg)	Time (hr)
80	51.0
70	24.8
60	16.9
50	13.1
40	10.9
30	9.7
20	8.9
10	8.5
0	8.4

hours per Mars year, while at the equator the available accumulation time is about 8 hours.

Table III gives the expected gamma-ray count rates for 13 elements. These results were calculated using the neutron and gamma ray transport codes used by Masarik and Reedy (1996) for the model composition given in the table. To first order, if different calculated count rates are needed for a different composition of Mars, the count rates will be proportional to the concentration of the element in question. For differences in concentration of elements like Fe, Cl, and H that have a large effect on the neutron flux distribution, the count rates cannot simply be scaled to obtain new results as discussed above. Nevertheless, the observed count rates and the resulting accumulation times required to measure the concentration of an element to 10% uncertainty in Table III can be used to calculate approximate sensitivities and uncertainties for other compositions and counting times as long as the composition is not greatly different from that used in the model. The sensitivities scale inversely with the square root of the counting time, and for lines that are small compared to the background, inversely with the concentration.

In order to get adequate statistics, spectra taken at different times and places need to be added together. Because of the narrow peaks, a few keV, relative to the full scale range of the spectrometer, 10,000 keV, small drifts in the gain or offset of the spectrometer will result in shifts that are significant compared to the width of the peaks. We need to correct for this gain drift so that we do not get an artificial peak broadening due to the addition of spectra taken at slightly different gains. Because the ∼ 2-hr orbital temperature variations occur on a time scale rapid compared to that needed to determine peak positions that would permit a determ-

TABLE III

Calculated accumulation times required to achieve 10% percision.

Element	Energy (keV)	Mode	Model Composition	Signal (c/s)	Continuum (c/s)	Time for 10% precision (hr)
H	2223	Capture	0.11%	0.0017	0.24	2400
O	6129	Inelastic	42.3%	0.0223	0.34	20
Mg	1369	Inelastic	5.2%	0.0124	0.37	70
Al	2210	Inelastic	4.2%	0.0029	0.24	820
Al	7724	Capture	4.2%	0.0008	0.25	12000
Si	1779	Inelastic	19.8%	0.0468	0.29	4
Si	3539	Capture	19.8%	0.0035	0.15	370
S	5424	Capture	2.7%	0.0021	0.37	2200
Cl	6111	Capture	0.55%	0.0081	0.34	150
K	1461	Radioactive	0.51%	0.1074	0.35	1
Ca	1943	Capture	4.7%	0.0018	0.27	2300
Mn	7244	Capture	0.4%	0.0009	0.28	9100
Fe	847	Inelastic	17.3%	0.0268	0.59	24
Fe	7632	Capture	17.3%	0.0130	0.26	44
Th	2614	Radioactive	0.30 ppm	0.0037	0.20	430
U	1765	Radioactive	0.078 ppm	0.0011	0.30	6800

ination of the gain and offset, we make corrections based on the temperature of the different components in the signal chain and laboratory determined coefficients of gain and offset variations with temperature.

For all but the strongest emission lines, it will be necessary to sum spectra not only over a long time period, but over different footprints. For example, if one wanted to determine an element such as aluminum to a precision of 10%, Table III indicates about 1000 hours are needed. Table II indicates that near the equator we only expect to get about 8 hours of data in each 450-km GS footprint. Thus it is necessary to degrade the spatial resolution to sum together spectra taken over different footprint locations to improve the statistics. Figure 9 shows the effect on a portion of the spectrum by increasing the accumulation time. In some cases the resolution can simply be degraded to a regular grid, e.g. 20-degree squares, or we can define larger regions, such as the highlands and the lowlands, based on other criteria and generate an average composition for those independently-defined regions.

Spectra that are collected over a large region, however, are more difficult to analyze, since the counts in the peaks depend on the thickness of the atmosphere, which attenuates the gamma rays, and the abundances of other elements that can

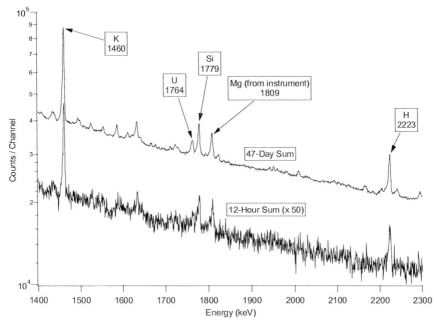

Figure 9. An expanded portion of the full-Mars GS spectrum shown in Figure 8 with a similar spectrum but with a much shorter accumulation time. The nature of the peak shapes can be seen; the area of the peak above the continuum is proportional to the concentration of the element responsible for the gamma-ray emission. The short-duration spectrum, collected for 12 hours, is what is expected for a 450-km footprint at middle latitudes. The uranium line is barely detectable in this spectrum, which shows the importance of being able to sum spectra together over larger regions to improve statistics for weak peaks.

moderate or absorb the neutrons. The surface composition will vary from place to place, and the atmospheric thickness will vary over both space and time based on elevation and season. It is thus difficult to solve the inverse problem of calculating the average composition based on the counts in the spectrum. Instead, we calculate the expected counts in the spectrum based on individual spatial elements, each having its own composition and atmospheric thickness, the latter including its time variation. The problem will be solved by iteration of the composition. Because the sensitivity is only weakly dependent on composition of other elements, we expect the solution to converge easily.

This method of forward calculation is required, as we already know that hydrogen is not uniformly distributed over the surface (Boynton *et al.*, 2002; Feldman *et al.*, 2002a; Mitrofanov *et al.*, 2002) having a much greater concentration near the poles and in some regions near the equator. As mentioned above, hydrogen has a very strong effect on neutron flux and this effect cannot be neglected.

4.4. REDUCTION OF NS NEUTRON DATA

The primary information needed from the NS to infer the hydrogen content and its stratigraphy in near-surface layers, or the thickness of carbon dioxide frost that covers the polar caps during winter is the amplitudes of thermal and epithermal fluxes that leak upward from Mars. Of course a more thorough determination of the reservoirs of martian volatiles and their stratigraphy requires a combined analysis of the 2.223 MeV hydrogen neutron capture line with the thermal and epithermal neutron counting rates as was done by Boynton et al. (2002) for the south polar region of Mars. Although possible for some types of neutron detectors, these amplitudes cannot be determined directly from the measured count rates of individual prism faces of NS. They can, however, be derived by combining the data from different faces as outlined below. A more complete description is given by Feldman et al. (1993b, 2002b).

4.4.1. Expected background and foreground count rates

The first stage of reduction involves a determination of counts that result from neutron capture by the boron contained in the scintillator prisms of the NS sensor. The spectrum of light output from the prisms resulting from background interactions is expected to be a broad continuum, dominated at low amplitudes by gamma rays and at high amplitudes by penetrating charged particles. In contrast, the spectrum resulting from low-energy neutron absorption will be peaked because of the high energy of the emitted α particle in the $^{10}B(n, \alpha)^7Li$ reaction relative to the energy of the absorbed neutron. An example of such a spectrum measured at Los Alamos National Laboratory using the flight detector is shown in Figure 4a. The peaked nature of this spectrum allows separation of neutron counts from background counts in spectra measured in Mars orbit through use of the measured calibration spectrum in Figure 4a and a least squares fitting routine. Whereas the background counts under the portion of the neutron peak that will be used in the fit is estimated to be about 5 s^{-1}, that due to planetary neutrons is estimated to range between 1 s^{-1} and 10 s^{-1}. The large range of estimated neutron count rates reflect different surface chemistries as well as differences in the orientation of the trapezoid faces; for example, the backward-looking face outruns thermal neutrons and so will register the fewest planetary neutron counts.

4.4.2. Adjustment for surface area response functions

Because the separate prisms of the NS have different orientations relative to the velocity vector of the Mars Odyssey spacecraft, they respond preferentially to neutrons evolving from different surface areas of Mars. This effect is shown for the forward and backward-facing prisms in Figure 10. Inspection shows that the response function of the forward-facing prism is offset from the nadir point in the direction of the velocity vector and that of the backward-facing prism is offset in the opposite direction. Thus the next step in the neutron data reduction chain must

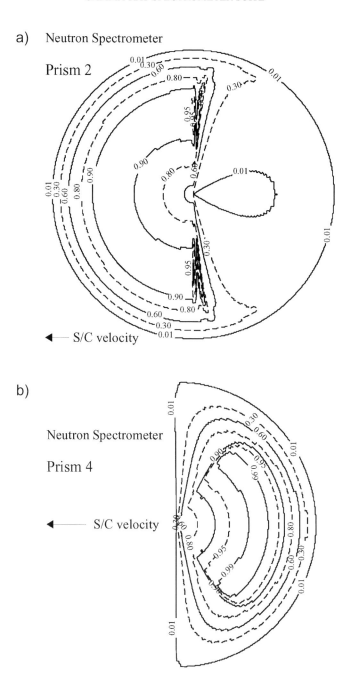

Figure 10. Relative efficiency of forward and rearward-facing prisms of the NS. The forward-facing prism detects neutrons mostly from the direction ahead of the spacecraft, while the rearward-facing detector sees only neutrons behind the spacecraft. The forward-facing detector has some probability of detecting neutrons emitted behind it, because some of those neutrons can overtake the spacecraft to be swept up by the forward prism.

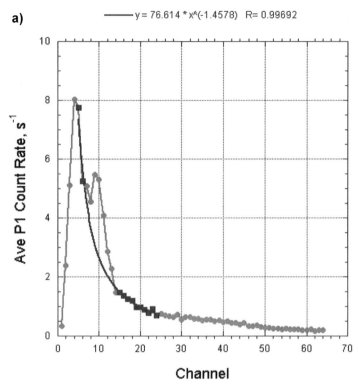

Figure 11. NS spectra of Mars showing how the net thermal neutron counts are removed from the continuum. (a) prism 1, nadir direction. (b) prism 2, velocity direction. (c) prism 3, zenith direction. (d) prism 4, anti-velocity direction.

be to rearrange the neutron counts measured by each of the sensor prisms during successive pixel intervals to fill the portions of the Martian surface overlain by their respective surface response functions.

A preliminary analysis of the response function of the Neutron Spectrometer and the angular emission of neutrons from the top of the Martian atmosphere shows that the size of the surface response functions, about 600 km full width at half maximum, is comparable to the sizes of the Martian polar caps. Because photomosaic maps of Mars from Viking show many other regional features that have sizes of this scale or smaller, we expect that neutron count rates may vary significantly as a function of orbital location. Variations in measured count rates will therefore not reflect actual variations in leakage neutron fluxes but will appear more muted. Inversion of the raw rates to obtain thermal and epithermal amplitudes using an algorithm based on simulations of a globally uniform chemistry will therefore incorrectly identify these amplitudes. A specific example of this effect, tailored to the measured configuration of the south polar cap of Mars, was presented by Feldman *et al.* (1993a). Minimization of errors therefore requires a deconvolution

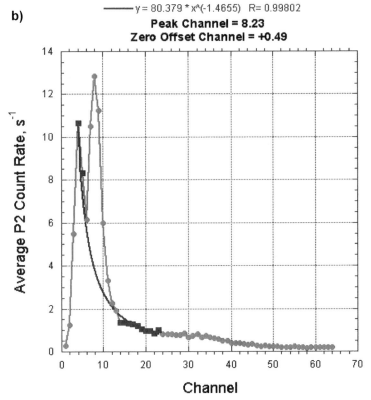

Figure 11. Continued.

of measured in-track, as well as cross-track, count rate variations before an accurate count rate inversion can be performed.

4.4.3. *Inversion of count rates to determine thermal and epithermal amplitudes*

Several studies of simulated neutron leakage fluxes from Mars have shown that resultant energy spectra can be fit well by a function of the form (e.g., Drake *et al.*, 1988; Dagge *et al.*, 1991, Feldman *et al.*, 2000).

$$F_m = \alpha(K/T)\exp(BK/T) + \beta(K/K_o)^{-p} F_c(K/K_o). \tag{1}$$

Here F_m is the model flux or current function, K is the neutron kinetic energy, α is the thermal amplitude, T is the temperature of the thermal distribution, β is the epithermal amplitude, p is the epithermal power law index, $F_c(K/K_o)$ is an arbitrary cutoff function to facilitate separation of thermal and epithermal components of F_m, and K_o is an arbitrary constant. Experimentation with many simulations showed that a cutoff function of the form

$$F_c(K/K_o) = [1 + (K/K_o)^{-k}]^{-1} \tag{2}$$

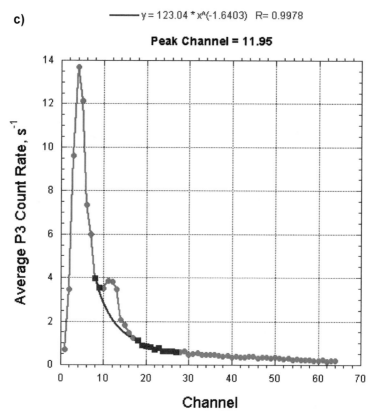

Figure 11. Continued.

provided both good fits to simulated spectra and clean thermal-epithermal separations with the choices of $K_o = 0.15$ eV and $k = 5$. Although this procedure has been revisited using results from Lunar Prospector (Feldman *et al.*, 2000), the new procedure is essentially the same as that just presented.

Count rates of the prisms of the NS sensor have been calculated for an extensive set of assumed Martian surface chemistries using a calculated energy-dependent efficiency function (Feldman *et al.*, 1993a). Best fit thermal and epithermal amplitudes, α and β from (1), were also calculated for each of the simulated neutron spectra. Intercomparison showed that the thermal amplitude is primarily related to the difference in count rates measured using the forward- and backward-facing prisms, and the epithermal amplitude can be determined by the count rates measured using both the downward- and backward-facing prisms. The number of neutrons from Mars that is reprocessed by the spacecraft can be continuously monitored using the upward-facing prism and removed using calibration data measured during the transition to the mapping orbit about Mars.

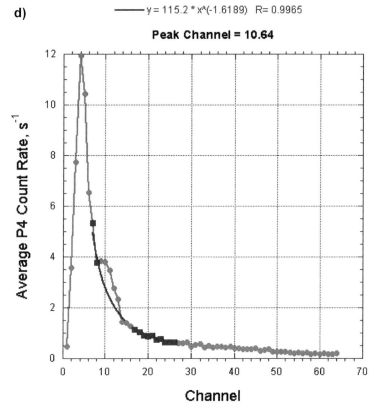

Figure 11. Continued.

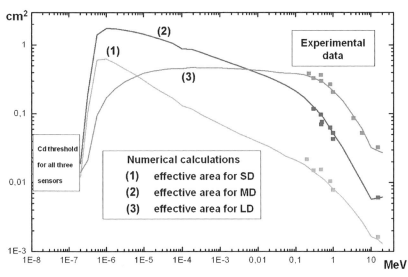

Figure 12. Calculated and measured efficiency (expressed as effective cross-sectional area) of the HEND neutron detectors. The low-energy cut off is due to the Cd filter.

4.5. ANALYSIS OF HEND NEUTRON DATA

HEND neutron data are presented in four signals S_{SD}, S_{MD}, S_{LD} and $S_{SC/IN/N}$, which all together cover the energy range from 0.4 eV up to 15 MeV. Four detectors of these signals SD, MD, LD and SC/IN/N have different dependence of cross-section on energy, and one may perform the spectral deconvolution of the detected neutron flux using the HEND calibration data (Figure 12). During the mission HEND data will be synchronized with pixels of the GS data, and all four HEND signals for neutrons will be accumulated in pixels for a global map of Mars.

The data from HEND will be combined with the data from the NS, which represent other energy ranges of neutrons. These two instruments provide the comprehensive description of Martian neutrons from thermal energy up to 15 MeV.

5. Science

The data obtained by the Mars Odyssey GRS will address a variety of scientific issues in several different areas of space sciences. The following sections describes the most important of these applications of the data from the GRS instrument suite.

5.1. GEOSCIENCE INVESTIGATIONS AT MARS

5.1.1. *Crust and mantle composition*

Processes of crustal formation on Mars are poorly known. Crustal formation is likely to have been complex; for example, in the most ancient cratered terrains, one sees clear evidence for nearly concurrent action of intense impact and volcanic processes. On other bodies in the inner solar system, crustal differentiation processes have concentrated some elements (e.g., Ca, Al) in crustal materials. We do not yet have a good understanding of the extent to which this has taken place on Mars. The abundances of refractory incompatible elements like Th and U should provide particularly useful information about the degree of fractionation of the martian crust with respect to the mantle. Th, U, and also K abundances in the martian crust are, of course, also important as constraints on the magnitude of radiogenic heating sources present, and K/U and K/Th ratios may be used to estimate the abundances of K and other moderately volatile elements in the planet.

Recent data from the Mars Global Surveyor Thermal Emission Spectrometer (TES) have provided insight into the mineralogy of martian crustal materials (e.g., Bandfield *et al.*, 2000). TES data show two primary mineralogical signatures in low-albedo regions on the martian surface. One is an apparently basaltic composition dominated by plagioclase and clinopyroxene, while the other appears more andesitic and is dominated by plagioclase and volcanic glass. The former is concentrated in the ancient cratered terrain of the southern highlands, while the latter predominates in the younger northern lowlands. While these data yield important mineralogical information, they provide no direct indication of chemistry. For the

Pathfinder landing site, APXS elemental analyses of rocks (e.g., Rieder *et al.*, 1997) have revealed a chemistry consistent with the andesite interpretation, and have placed some useful constraints on crustal evolution processes there, suggesting, for example, derivation from a high-alumina parent magma produced via early melting of a relatively primitive martian mantle (McSween *et al.*, 1999). The synergistic use of TES and GRS data together has the potential to further constrain the crustal composition inferences derived so far from the TES data, and (unlike the Pathfinder data) to do so on a global scale.

Determining the composition of the martian mantle is another major goal of geochemical investigation of the planet. The Mars Odyssey GRS may be able to perform analyses of volcanic materials derived from partial melting of the mantle, and from these to infer something about mantle composition. As one example, we can derive information about mantle chemistry from the concentration in volcanic rocks of elements, such as Fe and Mn, that do not fractionate substantially between solids and melts.

5.1.2. *Weathering processes*
Much of the martian surface appears to be dominated by fine-grained dust. The elemental chemistry of this material was investigated to first order by the Viking landers, and later by the Sojourner rover. The composition has been found to be rather similar at all three sites, suggesting that the materials at all of these locations are dominated by eolian deposits that have been largely homogenized on a global scale by dust storms. To the extent that compositional variations are observed in martian soils, they are explicable primarily as a result of varying admixtures of sulfate/chloride cement and local rock fragments (McSween and Keil, 2000).

The fines are clearly weathering products of some sort, but only a limited amount of information about the source rocks may be derived from our present knowledge of their elemental composition. Compared to most terrestrial materials, martian fines are high in Fe and Mg, and low in Si, Al, and K. These characteristics are consistent with weathering from mafic or ultramafic source rocks. However, our knowledge of dust chemistry is incomplete, and will be improved by the GRS. Moreover, it is not known to what extent the source rock composition implied by the chemistry of the fines is representative of martian surface rocks as a whole. For example, if the fines that are now globally distributed were generated primarily by local palagonitization of basaltic lavas via interaction with ground ice and water, the composition of the fines may be essentially unrelated to that of much of the planet. We will therefore attempt to compare the global average dust composition to the compositions of rocks determined at a range of sites.

It is clear that our signal will always represent some mixture of signal from both rock and fines. It was found at the Mars Pathfinder site that the fines were much lower in potassium than were the rocks (Wänke *et al.*, 2001). We may find some element, e.g. chlorine, which is enriched in fines that can be used as a tracer for the amount of fines. In cases where likely source rock chemistry can be identified, comparison of the compositions of source rocks and weathering products will yield insight into the weathering process. An element particularly important in this regard is hydrogen. Viking results suggested that some water of hydration was present in martian fines, but the amount was poorly determined. Results from the GRS should give the H concentration in the martian soil with high accuracy, indicating the degree of hydration that the source rocks have undergone.

At both of the Viking landing sites, near-surface fines were found to be cemented. Viking x-ray florescence data indicate that the cemented soil is higher in S and perhaps Cl than are the other materials analyzed. Pathfinder APXS measurements did not show any difference in S and Cl between the normal soils and a cemented deposit, called "Scoopy Doo" (Wänke et al., 2001). Therefore, it is still open as to how the cemented deposits were formed, since only their physical properties (hardness) seem to be different. Both soils and cemented deposits have high S and Cl concentrations. The S probably is present as sulfates; no sulfur-containing species were detected by the Viking GCMS experiment, as would be expected if the S was present in sulfides. The possible enhanced concentration of Cl in duricrust may indicate the presence of NaCl. The MO GRS should allow global investigation of the concentration of Cl and possibly S in the martian soil, permitting us to search for correlations of salt concentration with other geological indicators of the action of water.

5.1.3. *The SNC connection*

According to their trace element ratios and oxygen isotope characteristics, the SNC meteorites form a genetically related group of more than a dozen differentiated meteorites. The low crystallization ages and the trapped gases with element and isotope ratios of the martian atmosphere observed in one of them are considered to provide strong evidence for the proposition that the SNC meteorites represent martian surface rocks, ejected into space by large impacts. Based on the chemistry of the SNC meteorites and their mineralogy, precise information about the bulk composition (crust, mantle, and core) and structure of their assumed parent body Mars has been obtained (e.g., Wänke and Dreibus, 1988).

The GRS should further elucidate the connection between the SNC meteorites and Mars. Pairs of elements occurring globally in invariant abundance ratios, e.g. K/U or K/Th, will be most helpful in this respect. However the K/U ratio is quite variable in SNC meteorites and the preliminary value derived from the gamma-ray experiment of Phobos 2 falls outside the SNC range (Trombka *et al.*, 1992). Nevertheless, the K/U and K/Th ratios obtained by GRS will be of great importance to this question. In addition, the Fe/Mn ratios have potential importance. For all invariant ratios, integration over large areas or even over whole planet is possible, reducing the uncertainties by large factors. However, the possible shift of element ratios due to the weathering process may cause problems.

5.1.4. *Volcanism*

Volcanism has been a very important process in shaping the martian surface, and volcanic units are widespread on the planet. While the intensity of volcanic activity has generally diminished with time, volcanism appears to span almost all of recorded martian geologic history. Volcanic activity has been dominated throughout the planet's history by the extrusion of very fluid lavas, with yield strengths and viscosities (where they can be estimated) comparable to or lower than those of terrestrial basalts. In many instances, lavas have been emplaced as broad, flat volcanic plains. These plains range in age from the very old units interspersed through the ancient cratered terrain and on the floor of the Hellas basin, to the somewhat younger ridged plains of Lunae Planum, Chryse Planitia, and elsewhere, to the still younger plains of Elysium and, finally, Tharsis. It is not known to what extent there are systematic changes in the composition of these plains units with age, or what such changes might tell us about, say, changes in the depth of magma production with time. The Mars Odyssey GRS will be able to map the compositions of these units and to search for evidence of compositional changes.

Interspersed with and lying atop volcanic plains units, particularly in the Elysium and Tharsis regions, are Mars' spectacular shield volcanoes. These, too, span a considerable range in age, from the old, highly degraded shields of Hadriaca and Tyrrhena Paterae, through Alba Patera and the Elysium volcanoes, to the very young shields of Olympus Mons and on the Tharsis ridge. Again, little is known about variations in volcano composition with location and age. The largest martian volcanoes will be resolvable in the Odyssey GRS data, allowing us to search for variations in chemistry among them.

5.1.5. *Volatile reservoirs and transport*

The Mars Odyssey GRS experiment will also address a number of problems having to do with volatiles – primarily H_2O and CO_2 – at and below the martian surface. Both cosmochemical considerations and widespread evidence for fluvial activity indicate that a substantial amount of water was outgassed early in martian history. Much of that water may now reside beneath the ground as ice. Using simple models of heat and vapor transport in the martian regolith, it is possible to show that ground ice may be stable within tens of cm of the martian surface at latitudes within 30 to 40 degrees of the poles (e.g., Zent *et al.*, 1986). In fact, Mars Global Surveyor MOC images have shown strong evidence for recent localized melting of ground ice at some high-latitude locations (Malin and Edgett, 2000). Already the GRS has found evidence for subsurface ice at Mars, with a subsurface ice content of about 35% by mass buried by amounts of ice-free regolith that vary from 150 g/cm^2 to 30 g/cm^2 between −45 and −75 degrees latitude (Boynton *et al.*, 2002).

Overlying the layered deposits at both poles are deposits of perennial ice. The thickness of these deposits is poorly known, and it is not known to what extent their ice/dust ratio may differ from that of the underlying layered deposits. At the south pole, there is evidence from Viking that the perennial ice surface is CO_2 rather than H_2O (Kieffer, 1979), although this may not be the case in all years (Jakosky and Haberle, 1990). The perennial ice at both poles covers large enough areas to be resolved by the GRS, and it should be possible to determine the abundances of H_2O, CO_2, and dust accurately. Again, gamma-ray and neutron data may be used together to establish whether there are any vertical variations in composition over depths of tens of g/cm^2.

Overlying both poles in the winter and extending down to the middle latitudes are the seasonal frost caps. These caps are composed of CO_2 condensed from the atmosphere, and contain little H_2O (Kieffer et al., 2000). Using GRS data, it should be possible to make maps of polar cap thickness as a function of time through the martian year, observing the caps through their full cycle of advance and retreat. The topographic thickness of these caps has recently been measured by the Mars Global Surveyor MOLA instrument to be tens of cm (Smith et al., 2001). Their column density (in g/cm^2) may be determined both by using neutron data and by observing the changing attenuation of strong gamma-ray lines from the underlying regolith. These gamma and neutron data will therefore provide an independent measure of the mass distribution of CO_2 within the evolving cap. Combining these data on column density with the MOLA data on thickness, will yield the density of the CO_2 ice.

As noted above, Viking XRF (Clark et al., 1982) and Mars Pathfinder APXS (Rieder et al., 1997) analyses yielded high concentrations of sulfur and chlorine in the martian soil at all three landing sites. It seems unlikely that sulfur was added to the soil in the form of Mg- or Ca-sulfate because neither Mg or Ca is in excess when compared to the Shergotty meteorite, which otherwise matches the Viking soil very closely if normalized to SiO_2. It may be that sulfur was introduced via a gas-solid reaction of SO_2 or SO_3 to the material of the martian surface rocks comminuted by impacts.

In accordance with the volatile content of terrestrial magmas, H_2O vapor dominates in terrestrial volcanic gases, followed by CO_2 and SO_2. The H_2O content of SNC meteorites is more than one order of magnitude below that of comparable terrestrial rocks while sulfur is about equal. Hence if SNC meteorites are indeed representatives of martian crustal rocks, SO_2 and CO_2 could be the dominant volatile species on Mars. Gases from volcanic intrusions can be expected to migrate through the regolith toward the surface. In the absence of a strong greenhouse effect, only CO_2 would stay in gaseous form, while SO_2 gas would feed solid liquid SO_2 tables and only slowly find its way to the atmosphere where it would quickly be oxidized to SO_3 consuming H_2O.

Although the sensitivity of GRS for sulfur is limited (Table II), it can be hoped that the distribution of sulfur and chlorine can be studied on a global scale and some information can be gained on the thickness of the regolith containing sulfur and chlorine and on the total inventory of these two elements in the martian surface layers. The presence of sulfuric acid would, of course, place strong constraints on carbonate formation.

5.1.6. *Atmospheric processes*

The 16-g/cm^2-thick martian atmosphere will attenuate gamma rays that escape from the martian surface, especially those with the lowest energies (Metzger and Arnold, 1970). The thickness of the martian atmosphere can be determined by the differential attenuation of two gamma rays with different energies (Metzger, 1984; Metzger *et al.*, 1986a). The best sets of gamma rays for atmospheric-thickness studies will be those with well known ratios for their fluxes and the best spread in their attenuation coefficients, such as the 0.9112 and 2.6146 MeV gamma rays in the thorium decay chain. The same approach can be used to monitor the deposition and disappearance of polar caps over a surface of a different composition.

As the seasonal CO_2 frost forms on the winter pole of Mars, a significant fraction of the atmosphere condenses in the polar region. It is likely that this transport of atmosphere to the polar regions will carry the minor constituents of the atmosphere, Ar and N_2, with it to be concentrated in the polar regions as the CO_2 condenses, only to be replaced by more atmosphere containing more Ar and N_2. Depending on the mixing between the mid latitudes and polar regions, it is possible that a substantial enrichment of Ar and N_2 could be observed over the winter poles. This enrichment could be observed by the effect of thermal neutron attenuation caused by N or by gamma rays emitted by Ar or N.

5.2. ASTROPHYSICAL INVESTIGATIONS

Measurements of interest in the fields of Astrophysics, Space Physics and Solar Physics will be carried out as part of the mission utilizing the GRS. A few examples of the types of investigations planned will be described.

5.2.1. *Gamma-ray bursts*

One of the most puzzling and present mysteries in astrophysics is the question of the origin of gamma-ray bursts. Because these powerful and energetic cosmic bursts of gamma radiation occur at random times from random directions in the sky, their studies can be the best facilitated by non-oriented instruments with broad field of view. These gamma-ray instruments do not have angular resolution, and the only way that the source directions of the transient events can be accurately determined is through the timing of arrival of a transient at several widely separated detectors in space.

There are several instruments now in Earth orbit, e.g. KONUS/Wind and HETE-2.The Ulysses mission, launched in 1990 could be used as one distant interplanetary point. At least one more far-separated detector is necessary for accurate localization by the triangulation method, and GRS with HEND does provide this second interplanetary point.

The program of interplanetary triangulation of gamma-ray bursts has already been started by GRS and HEND during the Odyssey cruise flight. About 30 bursts were detected by HEND during 4.5 months of cruise, and several of them were localized with small error boxes on the sky (Hurley et al., 2002). This program will be continued during the mapping stage of Odyssey mission, and detection of about 24 new bursts per year could be expected from the Martian orbit.

5.2.2. Extragalactic gamma-ray background

The extragalactic gamma-ray background (EGB) is of great interest in cosmology and also is of potential significance for astrophysics. The first measurement of the EGB was obtained with a gamma-ray spectrometer aboard the Ranger spacecraft in 1961 (Metzger et al., 1964). Models for the origin of this radiation can be divided up into truly diffuse or very highly redshifted processes and source superposition models. The diffuse and redshifted (as opposed to source) production models involve cosmic-ray interaction, matter anti-matter annihilation, or exotic-particle decays. These processes are presumed to occur typically in the early history of the universe and involve the basic physical processes leading to the current form of the universe as we know it. It is for this reason that investigations of the low-energy gamma-ray background radiation studying both spectral and spatial characteristics will be of fundamental importance to our understanding of the universe. The observed low energy (1–10 MeV) gamma-ray background has been shown to have a general isotropy of at least 80% by NaI scintillator experiments (Trombka et al., 1977; Mazetz, Golenetskii and It'Inskii, 1977) and at least 90% by double-Compton telescope observations.

5.2.3. Solar processes

Impulsive energy release and high-energy particle acceleration often occur in cosmic plasmas at many sites throughout the Universe, ranging from planetary magnetospheres to accretion disks around compact objects. Nowhere can one pursue the study of this basic physics better than in the active Sun, where solar flares are the direct result of impulsive energy release and particle acceleration. Here, the acceleration of electrons is revealed by hard X-ray and gamma-ray bremsstrahlung; the acceleration of protons and nuclei is revealed by nuclear gamma-rays and neutrons. Understanding the processes that convert energy into high-energy particles is a major goal of astrophysics, and observations of solar flares offer a unique opportunity to study these processes. Gamma-ray line and hard X-ray continuum emissions have been observed from solar flares for many years and by many different instruments, yet measurements with the high-energy resolution necessary to extract all of the detailed information available from the gamma rays and hard X-rays are extremely limited.

Measurements of large Solar Particle Events (SPE) by the GRS will also be especially useful when the Mars Odyssey spacecraft is in position to view the hemisphere of the Sun not seen from the Earth. Such identifications will be very useful for global studies of traveling inter-planetary disturbances that affect cosmic-ray modulation. Also, flux measurements made by the Mars Odyssey GRS of energetic particles in the galactic cosmic rays and from the Sun will be compared with those from other spacecraft to help study the transport of these particles in the inner solar system.

Acknowledgements

The authors wish to thank the following people for their great help in supporting the design, fabrication and testing of the GRS instrument: S. Bailey, S. Battel, M. Berst, E. Boudreau, D. Burke, J. Crow, G. Davidson, H. Enos, D. Ferguson, M. Fitzgibbon, P. Gill, J. Goldsten, R. Hubble, S. Jung, M. Kennedy, W. J. Ko, D. Landis, C. Lansil, C. Lashley, B. Lawrie, L. Lebeau, R. Marcialis, S. Murphy, J. Odom, A. Post, L. Proctor, M. Rippa, R. Schmidt, D. Shepard, C. H. Song, N. Stevens, R. Stringfellow, M.-H. Tran, W. Verts, M. Williams, C. Wiswal.

References

Anders, E. and Ebihara, M.: 1982, 'Solar System Abundances of the Elements', *Geochim. Cosmochim. Acta* **46**, 2363–2380.

Arnold, J. R., Metzger, A. E., Anderson, E. C. and Van Dilla, M. A.: 1962, 'Gamma Rays in Space, Ranger 3', *J. Geophys. Res.* **67**, 4878–4880.

Arnold, J. R., Boynton, W. V., Englert, P. A. J., Feldman, W. C., Metzger, A. E., Reedy, R. C., Squyres, S. W., Trombka, J. I. and Wänke, H.: 1989, Scientific Considerations in the Design of the Mars Observer Gamma-Ray Spectrometer. In *High-energy Radiation Background in Space* (A. C. Rester, Jr. and J. I. Trombka, eds.), *AIP Conf. Proc.* **186**, 453–466.

Bandfield, J. L., Hamilton, V. E., and Christensen, P. R.: 2000, 'A Global View of Martian Surface Compositions from MGS-TES', *Science* **287**, 1626–1630.

Bielefeld, M. J., Reedy, R. C., Metzger, A. E., Trombka, J. I. and Arnold, J. R.: 1976, 'Surface Chemistry of Selected Lunar Regions', *Proc. Lunar Sci. Conf. 7th*, 2661–2676.

Boynton, W. V., Trombka, J. I., Feldman, W. C., Arnold, J. R., Englert, P. A. J., Metzger, A. E., Reedy, R. C., Squyres, S. W., Wanke, H., Bailey, S. H., Bruckner, J., Callas, J. L., Drake, D. M., Duke, P., Evans, L. G., Haines, E. L., McCloskey, F. C., Mills H., Shinohara, C. and Starr, R.: 1992, 'Science Application of the Mars Observer Gamma Ray Spectrometer', *JGR* **97**, 7681–7698.

Boynton, W. V., Evans, L. G., Starr, R., Brückner, J., Bailey, S. H. and Trombka, J. I.: 1998, 'Induced Backgrounds in the Mars Observer Gamma-Ray Spectrometer', in *Conference on the High Energy Radiation Background in Space*, IEEE Nuclear and Plasma Sciences Society, The Institute of Electrical and Electronic Engineers, Inc., Workshop Record 97TH8346, pp. 30–33.

Boynton, W. V., Feldman, W. C., Squyres, S. W., Prettyman, T. H., Bruckner, J., Evans, L. G., Reedy, R. C., Starr, R., Arnold, J. R., Drake, D. M., Englert, P. A. J., Metzger, A. E., Mitrofanov, Igor, Trombka, J. I., d'Uston, C., Wanke, H., Gasnault, O., Hamara, D. K., Janes, D. M., Marcialis, R. L., Maurice, S., Mikheeva, I., Taylor, G. J., Tokar, R. and Shinohara, C.: 2002, 'Distribution of Hydrogen in the Near Surface of Mars: Evidence for Subsurface Ice Deposits', *Science* **297**, 81–85.

Brückner, J., Koerfer, M., Wänke H., Schroeder, A. N. F., Filges D., Dragovitsch P., Englert P. A. J., Starr R., Trombka J., Taylor I., Drake D. and Shunk, E.: 1990, Radiation damage in germanium detectors: Implications for the gamma-ray spectrometer of the Mars Observer spacecraft. In: *Lunar and Planetary Science XXI* (Lunar and Planetary Institute, Houston), 137–138.

Brückner, J., Koerfer, M., Wänke, H., Schroeder, A. N. F., Filges, D., Dragovitsch, P., Englert, P. A. J., Starr, R., Trombka, J. I., Taylor, I., Drake, D. M. and Shunk, E. R.: 1991, 'Proton-Induced Radiation Damage in Germanium Detectors', *IEEE Transactions on Nuclear Science* **NS-38**, 209–217.

Brückner, J., Fabian, U., Patnaik, A., Wänke, H., Cloth, P., Dagge, G., Drüke, V., Filges, D., Englert,P. A. J., Drake, D. M., Reedy, R. C. and Parlier, B.: 1992, 'Simulation Experiments for Planetary Gamma-Ray Spectroscopy by Means of Thick Target High-Energy Proton Irradiations,' in *Lunar and Planetary Science XXIII* (Lunar and Planetary Institute, Houston), pp. 169–170.

Brückner, J., Wänke, H. and Reedy, R. C.: 1987, Neutron-Induced Gamma-Ray Spectroscopy: Simulations for Chemical Mapping of Planetary Surfaces. In Proceedings of the 17th Lunar and Planetary Science Conference, Part 2, *J. Geophys. Res.* **92**, B4, E603–E616.

Clark, B. C., Baird, A. K., Weldon, R. J., Tsusaki, D. M., Schnabel, L. and Candelaria, M. P.: 1982, 'Chemical Composition of Martian Fines', *J. Geophys. Res.* **87**, 10,059–10,067.

Dagge, G., Dragovitsch, P., Filges, D. and Brückner, J.: 1991, 'Monte Carlo Simulation of Martian Gamma-Ray Spectra Induced by Galactic Cosmic Rays', *Proc. Lunar Planet. Sci. Conf.* **21**, 425–435.

Drake, D. M., Feldman, W. C. and Jakosky, B. M.: 1988, 'Martian neutron leakage spectra', *J. Geophys. Res.* **93**, 6353–6368.

Etchegaray-Ramirez, M. I., Metzger, A. E, Haines, E. L. and Hawke, B. R.: 1983, 'Thorium concentrations in the lunar surface: IV. Deconvolution of the Mare Imbrium, Aristarchus, and adjacent regions', *J. Geophys. Res.* **88**, A529–A543.

Evans, L. G. and Squyres, S. W.: 1987, 'Investigation of Martian H_2O and CO_2 via orbital gamma-ray spectroscopy', *J. Geophys. Res.* **92**, 9153–9167.

Evans, L. G., Trombka, J. I. and Boynton, W. V.: 1986, 'Elemental analysis of a comet nucleus by passive gamma-ray spectrometry from a penetrator', *J. Geophys. Res.* **91**, B4, D525–D532.

Evans, L. G., Reedy, R. C. and Trombka, J. I.: 1993, Introduction to Planetary Remote Sensing Gamma Ray Spectroscopy, in *Remote Geochemical Analyses: Elemental and Mineralogical Composition* (C. M. Pieters and P. A. J. Englert, Eds.) (Cambridge Univ. Press, New York), pp. 167–198.

Evans, L. G., Trombka, J. I., Starr, R., Boynton, W. V. and Bailey, S. H.: 1998, Continuum Background in Space-Borne Gamma-Ray Detectors, in *Conference on the High Energy Radiation Background in Space*, IEEE Nuclear and Plasma Sciences Society, The Institute of Electrical and Electronic Engineers, Inc., Workshop Record 97TH8346, pp. 101–103.

Evans, L. D., Starr, R. D., Brückner, J., Reedy, R. C., Boynton, W. V., Trombka, J. I., Goldstein, J. O., Masarik, J., Nittler, L R. and McCoy, T. J.: 2001, 'Elemental composition from gamma-ray spectroscopy of the NEAR-Shoemaker landing site on 433 Eros', *Meteoritics Planetary Sci.* **36**, 1639–1660.

Feldman, W. C. and Drake, D. M.: 1986, 'A Doppler filter technique to measure the hydrogen content off planetary surfaces', *Nucl. Instrum. Methods Phys. Res.* **A245**, 182–190.

Feldman, W. C., Drake, D. M., O'Dell, R. D., Bringley, F. W., Jr. and Anderson, R. C.: 1989, 'Gravitational effects on planetary neutron flux spectra', *J. Geophys. Res.* **94**, 513–525.

Feldman, W. C. and Jakosky, B. M.: 1991, 'Detectability of martian carbonates from orbit using thermal neutrons', *J. Geophys. Res.* **96**, 15,589–15,598.

Feldman, W. C., Boynton, W. V., Jakosky, B. M. and Mellon, M. T.: 1993, 'Redistribution of subsurface neutrons caused by ground ice on Mars', *J. Geophys. Res.* **98**, #E11, 20855–20870.

Feldman, W. C., Boynton, W. V. and Drake, D. M.: 1993b, Planetary neutron spectroscopy from orbit, in Remote Geochemical Analysis: Elemental and Mineralogical Composition, C. M. Pieters, P. A. J. Englert, eds., pp. 213–234, Cambridge Univ. Press, New York.

Feldman, W. C., Barraclough, B. L., Fuller, K. R., Lawrence, D. J., Maurice, S., Miller, M. C., Prettyman, T. H. and Binder, A. B.: 1999, 'The Lunar Prospector Gamma-Ray and Neutron Spectrometers', *Nucl. Instr. Methods Phys. Res. A* **422**, 562–566.

Feldman, W. C., Lawrence, D. J., Elphic, R. C., Vaniman, D. T., Thomsen, D. R., Barraclough, B. L., Maurice S. and Binder, A. B.: 2000, 'Chemical information content of lunar thermal and epithermal neutrons', *J. Geophys. Res.* **105**, 20347–20363.

Feldman, W. C., Maurice, S., Lawrence, D. J., Little, R. C., Lawson, S. L., Gasnault, O., Wiens, R. C., Barraclough, B. L., Elphic, R. C., Prettyman, T. H., Steinberg, J. T and Binder, A. B.: 2001, 'Evidence for Water Ice Near the Lunar Poles', *J. Geophys. Res. Planets* **106**, #E10, 23231–23252.

Feldman, W. C., Boynton, W. V., Tokar, R. L., Prettyman, T. H., Gasnault, O., Squyres, S. W., Elphic, R. C., Lawrence, D. J., Lawson, S. L., Maurice, S., McKinney, G. W., Moore, K. R. and Reedy, R. C.: 2002a, 'Global Distribution of Neutrons from Mars: Results from Mars Odyssey', *Science* **297**, 75–78.

Feldman, W. C., Prettyman, T. H., Tokar, R. L., Boynton, W. V., Byrd, R. C., Fuller, K. R., Gasnault, O., Longmire, J. L., Olsher, R. H., Storms, S. A. and Thornton, G. W.: 2002b, 'Fast neutron flux spectrum aboard Mars Odyssey during cruise', *J. Geophys. Res.* **107**, 10.1029/2001JA000295.

Fermi, E.: 1950, Nuclear Physics, Univ. Chicago Press, p. 248.

Gasnault , O., Feldman, W. C., Maurice, S., Genetay, I., d'Uston, C., Prettyman, T. H. and Moore, K. R.: 2001, 'Composition from Fast Neutrons: Application to the Moon', *Geophys. Res. Lett.* **28**, 3797–3800

Jakosky, B. M. and Haberle, R. M.: 1990, 'Year-to-year instability of the Mars south polar cap', *J. Geophys. Res.* **95**, 1359–1365.

Kieffer, H. H.: 1979, 'Mars south polar spring and summer temperatures: A residual CO_2 frost', *J. Geophys. Res.* **84**, 8263–8288.

Kieffer, H. H., Titus, T. N., Mullins, K. F. and Christensen, P. R.: 2000, 'Mars south polar spring and summer behavior observed by TES: Seasonal cap evolution controlled by frost grain size', *J. Geophys. Res.* **105**, 9653–9699.

Lapides, J. R.: 1981, Planetary gamma-ray spectroscopy: The effects of hydrogen and the macroscopic thermal-neutron absorption cross section on the gamma-ray spectrum. Thesis, University of Maryland, College Park, 115 pp.

Lawrence, D. J., Feldman, W. C., Barraclough, B. L., Elphic, R. C., Maurice, S., Binder, A. B., Miller, M. C. and Prettyman, T. H;: 1999, 'High Resolution Measurements of Absolute Thorium Abundances on the Lunar Surface', *Geophys. Res. Lett.* **26**, No. 17, 2681–2684.

Lawrence, D. J., Feldman, W. C., Elphic, R. C., Little, R. C., Prettyman, T. H., Maurice, S., Lucey, P. G. and Binder, A. B.: 2002, ' Iron abundances on the lunar surface as measured by the Lunar Prospector gamma-ray and neutron spectrometers', *J. Geophys. Res. Planets*, in press.

Lingenfelter, R. E., Canfield, E. H. and Hampel, V. E.: 1972, 'The lunar neutron flux revisited', *Earth Planet. Sci. Lett.* **16**, 355–369.

Lingenfelter, R. E., Canfield, E. H. and Hess, W. N.: 1961, 'The lunar neutron flux', *J. Geophys. Res.* **66**, 2665–2671.

Mahoney, W. A., Ling, J. C., Jacobson, A. S. and Tapphorn, R. M.: 1980, 'The HEAO 3 gamma-ray spectrometer', *Nucl. Instrum. Methods* **178**, 363–381.

Malin, M. C. and Edgett, K. E.: 2000, 'Evidence for recent groundwater seepage and surface runoff on Mars', *Science* **288**, 2330–2335.

Masarik, J. and Reedy, R. C.: 1994, 'Effects of Bulk Composition on Nuclide Production Processes in Meteorites', *Geochim. Cosmochim. Acta* **58**, 5307–5317.

Masarik, J. and Reedy, R. C.: 1996, 'Gamma Ray Production and Transport in Mars', *J. Geophys. Res.* **101**, 18,891–18,912.

Mazets, E. P., Golenetskii, S. V. and Il'Inski, V. N.: 1977, P.s'ma V Astron. Zh (USSR), Vol. 2, No. 12, 563.

McSween, H. Y., Murchie, S. L., Crisp, J. A., Bridges, N. T., Anderson, R. C., Bell, J. F. III, Britt, D. T., Brückner, J., Dreibus, G., Economou, T., Ghosh, A., Golombek, M. P., Greenwood, J. P., Johnson, J. R., Moore, H. J., Morris, R. V., Parker, T. J., Rieder, R., Singer, R. and Wänke, H.: 1999, 'Chemical, multispectral, and textural constraints on the composition and origin of rocks at the Mars Pathfinder landing site', *J. Geophys. Res.* **104**, 8679–8715.

McSween, H. Y. and Keil, K.: 2000, 'Mixing relationships in the Martian regolith and the composition of globally homogeneous dust', *Geochim. Cosmochim. Acta* **64**, 2155–2166.

Metzger, A. E.: 1984, 'Climatology capabilities of a gamma-ray spectrometer at Mars', *Bull. Am. Astron. Soc.* **16**, 678–679.

Metzger, A. E., Anderson E. C., Van Dilla M. A. and Arnold, J. R.: 1964, 'Detection of an interstellar flux of gamma-rays', *Nature* **204**, 766–767.

Metzger, A. E. and Arnold, J.R.: 1970, 'Gamma-ray spectroscopic measurements of Mars', *Appl. Opt.* **9**, 1289–1303.

Metzger, A. E., Arnold, J. R., Reedy, R. C., Trombka, J. I. and Haines, E. L.: 1986a, The application of gamma-ray spectroscopy to the climatology of Mars. In: *Lunar and Planetary Science XVII* (Lunar and Planetary Institute, Houston), 549–550.

Metzger, A. E. and Drake, D. M.: 1990, 'Identification of lunar rock types and search for polar ice gamma ray spectroscopy', *J. Geophys. Res.* **95**, 449–460.

Metzger, A. E. and Haines, E. L.: 1990, 'Atmospheric measurements at Mars via gamma ray spectroscopy', *J. Geophys. Res.* **95**, 14,695–14,715.

Metzger, A. E., Parker, R. H., Arnold, J. R., Reedy, R. C. and Trombka, J. I.: 1975, 'Preliminary design and performance of an advanced gamma-ray spectrometer for future orbiter missions', *Proc. Lunar Sci. Conf. 6th*, 2769–2784.

Metzger, A. E., Parker, R. H. and Yellin, J.: 1986b, 'High energy irradiations simulating cosmic-ray-induced planetary gamma ray production. I. Fe target', *J. Geophys. Res.* **91**, D495–D504.

Mitrofanov, I., Anfimov, D., Kozyrev, A., Litvak, M., Sanin, A., Tret'yakov, V., Krylov, A., Shvetsov, V., Boynton, W., Shinohara, C., Hamara, D. and Saunders, R. S.: 2002, 'Maps of Subsurface Hydrogen from the High Energy Neutron Detector, Mars Odyssey', *Science* **297**, 78–81.

Pehl, R. H., Varnell, L. S. and Metzger, A. E.: 1978, 'High-energy proton radiation damage of high-purity germanium detectors', *IEEE Trans. Nucl. Sci.* **NS-25**, 409–417.

Prettyman, T. H., Feldman, W. C., Lawrence, D. J., McKinney, G. W., Binder, A. B., Elphic, R. C., Gasnault, O. M., Maurice, S. and Moore, K. R.: 2002, 'Library least squares analysis of Lunar Prospector Gamma-ray spectra', *33rd Lunar and Planetary Science Conference, Abstract #2012*.

Reedy, R. C.: 1978, 'Planetary gamma-ray spectroscopy', *Proc. Lunar Planet. Sci. Conf. 9th*, 2961–2984.

Reedy, R. C.: 1988, Gamma-ray and neutron spectroscopy of planetary surfaces and atmospheres. In: *Nuclear Spectroscopy of Astrophysical Sources* (N. Gehrels and G. Share, eds.), AIP Conf. Proc. 170 (American Institute of Physics, New York), 203–210.

Reedy, R. C. and Arnold, J. R.: 1972, 'Interaction of solar and galactic cosmic-ray particles with the Moon', *J. Geophys. Res.* **77**, 537–555.

Reedy, R. C., Arnold, J. R. and Trombka, J. I.: 1973, 'Expected gamma ray emission from the lunar surface as a function of chemical composition', *J. Geophys. Res.* **78**, 5847–5866.

Rieder, R., Economou, T. Wänke, H., Turkevich, A., Crisp, J., Brückner, J., Dreibus, G. and Mc-Sween, H. Y. Jr.: 1997, 'The chemical composition of Martian soil and rocks returned by the mobile alpha proton X-ray spectrometer: Preliminary results from the X-ray mode', *Science* **278**, 1771–1774.

Saunders, R. S., Arvidson, R. E., Badhwar, G. D., Boynton, W. V., Christensen, P., Cucinotta, F. A., Gibbs, R. G., Kloss, Jr. C., Landano, M. R., Mase, R. A., Meyer, M., Pace, G., Plaut, J. J., Sidney, W., McSmith, G. W., Spencer, D. A., Thompson, T. W. and Zeitlin, C. J.: 2004, '2001 Mars Odyssey Mission Summary', *Space Sci. Rev.*, **110**, 1-36.

Smith, D. E., Zuber, M. T., and Neumann, G. A.: 2001, 'Seasonal variations of snow depth on Mars', *Science* **294**, 2141–2146.

Surkov, Y. A.: 1984, 'Nuclear-physical methods of analysis in lunar and planetary investigations', *Isotopenpraxis* **20**, 321–329.

Surkov, Y. A., Barsukov, V. L., Moskaleva, L. P., Kharyukova, V. P., Zaitseva, S. Y., Smirnov, G. G. and Manvelyan, O. S.: 1989, 'Determination of the elemental composition of martian rocks from Phobos 2', *Nature* **341**, 595–598.

Thakur, A. N.: 1997, 'Analysis Of Gamma-Ray Continuum Spectra to Determine the Chemical Composition', *J. Radioanal. Nucl. Chem.* **215**, 161–167.

Trombka, J. I., Dyer, C. S., Evans, L. G., Bielefeld, M. J., Seltzer, S. M. and Metzger, A. E.: 1977, 'Reanalysis of the Apollo Cosmic Gamma-Ray Spectrum in the 0.3 to 10 MeV Energy Region', *Astrophys. J.* **212**, 925–935.

Trombka, J. I., Evans, L. G., Starr, R., Floyd, S. R., Squyres, S. W., Whelan, J. T., Barnford, G. J., Coldwell, R. L., Rester, A. C., Surkov, Y. A., Moskaleva, L. P., Kharyukova, V. P., Manvelyan, O. S., Zaitseva, S. Y. and Smirnov, G. G.: 1992, 'Analysis of Phobos Mission Gamma-Ray Spectra from Mars', *Proc. Lunar Planet. Sci. Conf.* **22**, 22-39.

Trombka, J. I., Squyres, S. W., Brückner, J., Boynton, W. V., Reedy, R. C., McCoy, T. J., Gorenstein, P., Evans, L. G., Arnold, J. R., Starr, R. D., Nittler, L. R., Murphy, M. E., Mikheeva, I., McNutt Jr., R. L., McClanahan, T. P., McCartney, E. Goldsten, J. O., Gold, R. E., Floyd, S. R., Clark, P. E., Burbine, T. H., Bhangoo, J. S., Bailey, S. H. and Petaev, M.: 2000, 'The Elemental Composition of Asteroid 433 Eros: Results of the NEAR-Shoemaker X-ray Spectrometer', *Science* **289**, 2101–2105.

Van Dilla, M. A., Anderson, E. C., Metzger, A. E. and Schuch, R. .L.: 1962, 'Lunar composition by scintillation spectroscopy', *IRE Trans. Nucl. Sci.* **NS-9**, 405–412.

Wänke, H. and Dreibus, G.: 1988, 'Chemical Composition and Accretion History of Terrestrial Planets', *Phil. Trans. R. Soc. Lond. A* **325**, 545–557.

Wänke H., Brückner J., Dreibus G., Rieder R. and Ryabchikov I.: 2001, Chemical composition of rocks and soils at the Pathfinder site, Space Science Reviews, 96, 317-330.

Yadav, J. S., Brückner, J. and Arnold, J. R.: 1989, 'Weak Peak Problem in High Resolution Gamma-Ray Spectroscopy', *Nucl. Instrum. Methods Phys. Res.* **A277**, 591–598.

Zent, A. P., Fanale, F. P., Salvail, J. R. and Postawko, S. E.: 1986, 'Distribution and state of H_2O in the high-latitude shallow subsurface of Mars', *Icarus* **67**, 19–36.

THE THERMAL EMISSION IMAGING SYSTEM (THEMIS) FOR THE MARS 2001 ODYSSEY MISSION

PHILIP R. CHRISTENSEN[1*], BRUCE M. JAKOSKY[2], HUGH H. KIEFFER[3],
MICHAEL C. MALIN[4], HARRY Y. MCSWEEN, JR.[5], KENNETH NEALSON[6],
GREG L. MEHALL[7], STEVEN H. SILVERMAN[8], STEVEN FERRY[8],
MICHAEL CAPLINGER[4] and MICHAEL RAVINE[4]

[1] *Department of Geological Sciences, Arizona State University, Tempe, AZ 85287-1404, U.S.A.*
[2] *Laboratory of Atmospheric and space Physics and Department of Geological Sciences, University of Colorado, Boulder, CO 80309, U.S.A.*
[3] *U.S. Geological Survey, Flagstaff, AZ 86001, U.S.A.*
[4] *Malin Space Science Systems, San Diego, CA 92121, U.S.A.*
[5] *Department of Geological Sciences, University of Tennessee, Knoxville, TN 37996-1410, U.S.A.*
[6] *Jet Propulsion Laboratory, Pasadena, CA 91006, U.S.A.*
[7] *Department of Geological Sciences, Arizona State University, Tempe, AZ 85287-1404, U.S.A.*
[8] *Raytheon Santa Barbara Remote Sensing, Goleta, CA, U.S.A.*
(*Author for correspondence, E-mail: phil.christensen@asu.edu)

(Received 4 August 2001; Accepted in final form 6 December 2002)

Abstract. The Thermal Emission Imaging System (THEMIS) on 2001 Mars Odyssey will investigate the surface mineralogy and physical properties of Mars using multi-spectral thermal-infrared images in nine wavelengths centered from 6.8 to 14.9 μm, and visible/near-infrared images in five bands centered from 0.42 to 0.86 μm. THEMIS will map the entire planet in both day and night multi-spectral infrared images at 100-m per pixel resolution, 60% of the planet in one-band visible images at 18-m per pixel, and several percent of the planet in 5-band visible color. Most geologic materials, including carbonates, silicates, sulfates, phosphates, and hydroxides have strong fundamental vibrational absorption bands in the thermal-infrared spectral region that provide diagnostic information on mineral composition. The ability to identify a wide range of minerals allows key aqueous minerals, such as carbonates and hydrothermal silica, to be placed into their proper geologic context. The specific objectives of this investigation are to: (1) determine the mineralogy and petrology of localized deposits associated with hydrothermal or sub-aqueous environments, and to identify future landing sites likely to represent these environments; (2) search for thermal anomalies associated with active sub-surface hydrothermal systems; (3) study small-scale geologic processes and landing site characteristics using morphologic and thermophysical properties; and (4) investigate polar cap processes at all seasons. THEMIS follows the Mars Global Surveyor Thermal Emission Spectrometer (TES) and Mars Orbiter Camera (MOC) experiments, providing substantially higher spatial resolution IR multi-spectral images to complement TES hyperspectral (143-band) global mapping, and regional visible imaging at scales intermediate between the Viking and MOC cameras.

The THEMIS uses an uncooled microbolometer detector array for the IR focal plane. The optics consists of all-reflective, three-mirror anastigmat telescope with a 12-cm effective aperture and a speed of f/1.6. The IR and visible cameras share the optics and housing, but have independent power and data interfaces to the spacecraft. The IR focal plane has 320 cross-track pixels and 240 downtrack pixels covered by 10 $\sim$1-μm-bandwidth strip filters in nine different wavelengths. The visible camera has a 1024 × 1024 pixel array with 5 filters. The instrument weighs 11.2 kg, is 29 cm by 37 cm by 55 cm in size, and consumes an orbital average power of 14 W.

1. Introduction

A major goal of the Mars Exploration Program is to obtain data that will help determine whether life ever existed on Mars. This goal will ultimately be addressed via detailed *in situ* studies and eventual return to the Earth of samples of the martian surface. It is therefore essential to identify future landing sites with the highest probability of containing samples indicative of early pre-biotic or biotic environments. Of particular interest are aqueous and/or hydrothermal environments in which life could have existed, or regions of current near-surface water and/or heat sources [*Exobiology Working Group*, 1995]. The search for these environments requires detailed geologic mapping of key sites and accurate interpretations of the composition and history of these sites in a global context.

The Thermal Emission Imaging System (THEMIS) will contribute to these goals through the global mapping of unique compositional units and the identification of key minerals and rock types. This mapping will be accomplished at spatial scales that permit mineral and rock distributions to be related to the geologic processes and history of Mars. The specific objectives of the THEMIS investigation are to: (1) determine the mineralogy of localized deposits associated with hydrothermal or sub-aqueous environments, and to identify future landing sites likely to represent these environments; (2) search for pre-dawn thermal anomalies associated with active sub-surface hydrothermal systems; (3) study small-scale geologic processes and landing site characteristics using morphologic and thermophysical properties; (4) investigate polar cap processes at all seasons using infrared observations at high spatial resolution; and (5) provide a direct link to the global hyperspectral mineral mapping from the Mars Global Surveyor (MGS) Thermal Emission Spectrometer (TES) investigation by utilizing a significant portion of the infrared spectral region at high (100 m per pixel) spatial resolution.

These objectives will be addressed using thermal infrared (6.3–15.3 μm) multi-spectral observations in nine wavelengths at 100-m per pixel spatial resolution, together with 18-m per pixel visible imagery in up to five colors. The thermal-infrared spectral region was selected for mineral mapping because virtually all geologic materials, including carbonates, hydrothermal silica, sulfates, phosphates, hydroxides, and silicates have fundamental infrared absorption bands that are diagnostic of mineral composition. THEMIS visible imaging provides regional coverage (with global coverage a goal) at spatial scales that are intermediate between those of Viking and the detailed views from the MGS Mars Orbiter Camera (MOC) (Malin and Edgett, 2001).

THEMIS builds upon a wealth of data form previous experiments, including the Mariner 6/7 Infrared Spectrometer (Pimentel *et al.*, 1974) the Mariner 9 Infrared Interferometer Spectrometer (IRIS) (Hanel *et al.*, 1972; Conrath *et al.*, 1973), the Viking Infrared Thermal Mapper (IRTM) (Kieffer *et al.*, 1977), the Phobos Termoscan (Selivanov *et al.*, 1989), the MGS TES (Christensen *et al.*, 2001a; Smith *et al.*, 2001c; Bandfield *et al.*, 2000a; Christensen *et al.*, 2000b, 2001b), and the MGS

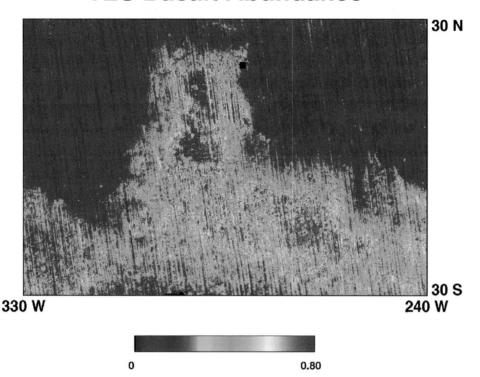

Figure 1. TES mineral map of the Syrtis Major region of Mars. The fraction of the surface covered by basalt is shown as derived from deconvolution of TES spectra. Data are from Bandfield *et al.* (2000). (a) Regional view. Black region is area shown in Figure 1b. (b) 60 km × 60 km region at TES resolution.

MOC (Malin and Edgett, 2001). In particular, the TES instrument has collected hyperspectral images (143 and 286 spectral bands) of the entire martian surface at wavelengths from 6 to 50 micrometers, providing an initial global reconnaissance of mineralogy and thermophysical properties (Christensen *et al.*, 2000b, d, 2001a; Bandfield *et al.*, 2000a; Bandfield, 2002; Ruff and Christensen, 2002; Jakosky *et al.*, 2000; Mellon *et al.*, 2000). By covering the key 6.3 to 15.0 μm region in both instruments it is possible to combine the high spectral resolution of TES with the high spatial resolution of THEMIS to achieve the goals of a global mineralogic inventory at the spatial scales necessary for detailed geologic studies within the Odyssey data resources.

Figure 1 shows a basalt abundance map derived from TES data of a region of Mars centered on Syrtis Major, illustrating the global-scale reconnaissance obtained from TES. Figure 1b shows a 60 km × 60 km region within Syrtis Major. Figure 2a shows an orbital infrared image of a 60 by 60 km area on Earth, degraded to 3 km per pixel. Figure 2b shows the same scene at 100-m per pixel THEMIS

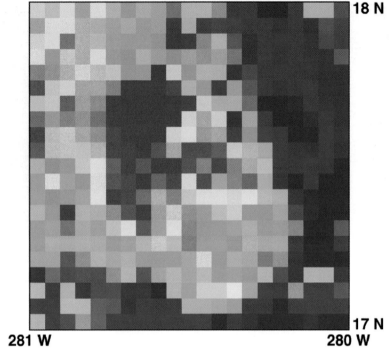

Figure 1. Continued.

resolution. Comparison of scenes at TES and THEMIS resolutions illustrates the dramatic improvement in geologic unit discrimination that is possible at THEMIS spatial resolution, and the difficulties encountered in attempting detailed studies for geologic interpretation and landing site selection using TES data alone. At THEMIS scale, terrestrial lake beds, hydrothermally altered zones, carbonate outcrops, silica-rich rocks, channel floors, and other indications of ancient aqueous environments can be readily identified and mapped using multi-spectral IR images. THEMIS can provide similar detection of such systems on Mars, if they are present at similar scales and exposed at the surface.

THEMIS data will be used to identify and map sites for future rovers and sample- return missions, beginning with the Mars 2003 Rovers, by aiding in the evaluation of the science rationale, hazards (e.g. rocks and dust), and morphology of these sites. These data will also provide information to adjust the exact location of the landing ellipses to maximize the science return, plan the '03 Rover traverses, and aid in the extrapolation of the Rover results. The Rover payloads contain miniature versions of the MGS TES instrument (Silverman *et al.*, 1999) covering the THEMIS spectral region. If both the Miniature-TES and THEMIS (and possibly TES) instruments are operating simultaneously, there will be a powerful link from

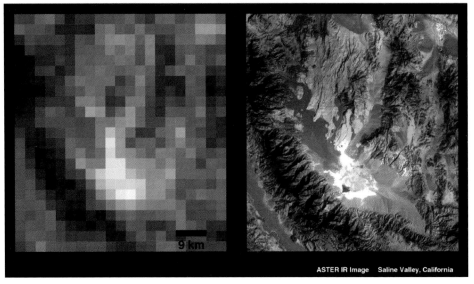

Figure 2. Comparison of TES and THEMIS spatial resolution. Both image sets are IR 3-band visible images of the Saline Valley, California acquired by the ASTER instrument on the Terra spacecraft. ASTER image. (a) Simulated 3-km per pixel TES resolution. (b) THEMIS 100-m per pixel resolution. Individual channels, lake beds, silica-rich outcrops and sediments, and lava flows can be spatially distinguished and compositionally identified. These data have been made available by NASA, GSFC, MITI, ERSDAC, JAROS, and the U.S./Japan ASTER Science Team.

the meter scale viewed by mini-TES to the 100-m scale of THEMIS to the global mapping of MGS TES.

2. Mars Science Questions

A primary objective of the Mars Exploration Program is to understand the biotic and pre-biotic environments of Mars. Key questions can be divided into two broad categories: (1) the search for aqueous environments, both ancient and active, that provide insight into biotic and pre-biotic conditions; and (2) the understanding changes in environmental conditions, climate, and geologic processes through time.

2.1. AQUEOUS ENVIRONMENTS

A major scientific focus of the Mars Program is the exploration of Mars as a possible abode for life. This exploration will be done by the global reconnaissance of Mars to understand the geological and volatile history, the detailed investigation of sites of exobiological interest, and the in-depth *in situ* exploration of some of these sites using rovers, culminating in the return of samples from one or more sites.

Although there is much that we do not know about the origin of life, the necessary environmental conditions appear to consist of: (i) the presence of liquid water, as a medium of transport for nutrients and waste products; (ii) the presence of the biogenic elements, consisting of C, H, O, N, S, P, Ca, Fe, and other trace elements that participate in life; and (iii) a source of energy that can drive chemical disequilibrium, so that the 'slide' back towards equilibrium can drive biochemical reactions (e.g., Jakosky, 1998). Mars appears to have (or have had) all of these ingredients:

(i) Evidence for liquid water is widespread on Mars (e.g. Carr, 1996). The ancient, heavily cratered surfaces (older than about 3.5 b.y.) are dissected by valley networks; the role of liquid water in their formation is clear, even though the exact mechanism of formation is being debated (Malin and Carr, 1999). In addition, there has been dramatic sedimentation and erosion on these surfaces, evidenced by the presence of extensive layered deposits (Malin and Edgett, 2000), the distribution of small impact craters, and the severe degradation of larger craters (e.g. Craddock and Maxwell, 1993; Craddock *et al.*, 1997). The high deposition and erosion rates, the catastrophic outflow channels, and the presence of morphological features such as layers and gullies in partly eroded craters, can best be explained by liquid water (e.g. Carr, 1996). Apparently, water was more stable and abundant near the surface of early Mars than it is today.

(ii) Mars appears to have all of the biogenic elements readily accessible at the surface or in the crust. This is not surprising given the active geological environment that has existed for 4 b.y. In particular, carbon is present in the atmosphere in a readily usable form (CO_2) that also can dissolve in water and percolate into the crust. In addition, *in situ* measurements from the Viking and Pathfinder landers, together with studies of the martian meteorites, indicate the presence of all of the other elements necessary to support life (e.g. McSween, 1994; Bell *et al.*, 2000).

(iii) Volcanism has occurred on Mars throughout time, and can provide a ready source of easily accessed geothermal energy. On Earth, volcanic heat is tapped by the circulation of water through hydrothermal systems. As the water is released back to the surface, chemical potential drives the formation of organic molecules that can serve as a reducing agent to do useful work (Shock, 1997). Such hydrothermal systems may have served as the location for the terrestrial origin of life, and may have been widespread on Mars given the abundant volcanism and crustal water (Jakosky and Shock, 1998).

Among the most likely sites to search for life are regions where liquid water was present for substantial periods of time. A great deal of work has been done to define the characteristics of such sites and to identify the environments that are conducive both to the sustenance of life and to the preservation of evidence of this life (McKay *et al.*, 1996; Boston *et al.*, 1992; Walter and Des Marais, 1993). Lacustrine sediments have been identified as logical targets both as evidence of an aqueous environment and because fossils are often preserved there. A second high priority target would be ancient thermal springs, where life could have existed and be well preserved in unique, remotely identifiable mineral deposits (Walter and Des Marais, 1993).

The characteristics of terrestrial spring deposits provide a context in which to search for similar martian deposits. These deposits derive from the intense hydrothermal weathering of the country rock. Silica, in the form of amorphous silica, quartz, chalcedony, and opaline silica, is the most common precipitate formed in volcanic terrains (Ellis and McMahon, 1977), due to both the high solubility of silica in hot water and the rapid decrease in solubility with decreasing temperature. Silica is also an excellent preservation media and is an ideal candidate site in the search for ancient life on Mars (Walter and Des Marais, 1993). Silica-depositing springs are abundant in a wide range of volcanic compositions on Earth, occurring in both rhyolitic (e.g., Yellowstone) and basaltic (e.g., Iceland) regions (Brock, 1978). The Yellowstone region is an excellent example of a THEMIS target. It contains $\sim$3,000 springs spread over an 80 by 100 km area, with zones of intense hydrothermal alteration up to several km in size (White et al., 1988).

Calcium carbonate deposits (travertine) are also common in hydrothermal systems. Hydrothermal travertine forms by degassing of hot ($\sim$60–100 °C) Ca- and CO_2-rich groundwater, followed by precipitation of calcium carbonate (Ellis and McMahon, 1977; Pentecost, 1996). Studies of travertine have indicated the common presence of organic material, including microfossils (Pentecost, 1996; Bargar, 1978).

An additional location to search for evidence of life would be deep beneath the surface, where martian biota might survive (and perhaps even thrive) in much the same way that some deep-subsurface bacteria thrive on the Earth–by metabolizing hydrogen produced by interactions between water and basalt in the pore spaces of the rock (Stevens and McKinley, 1995). Such locales on Mars might be observable in the walls of Valles Marineris, for example, where there is up to 10 km of vertical exposure of rock that were previously buried.

2.2. SURFACE MORPHOLOGY AND THE EVOLUTION OF CLIMATE

2.2.1. *Aqueous Environments and Morphology*

Aqueous environments can be distinguished by a variety of unique morphologic features. These include flow indicators, such as streamlined depositional or erosion features, layered sediments, strand lines, and delta deposits. These features can be detected with 18 m resolution, as demonstrated by the correlation between predictions for the Pathfinder site based on 38 m Viking Orbiter imagery (Golombek et al., 1997) and the actual Pathfinder surface observations. Paleohydrology will be studied using images of the sequence of sediments deposited by various stages of fluvial processes. These will include variations in flow hydraulics, sediment transport characteristics, and the properties and distribution of deposits (Komatsu and Baker, 1997). The visible images also provide excellent context images for the 100-m multi-spectral IR and nighttime temperature images, allowing the distribution of compositionally unique outcrops, deposits, or disseminated materials to be interpreted in relationship to the landscape processes.

2.2.2. Aeolian Sediments and Morphology

Sedimentary deposits on Mars range from micron thick dust coatings to km-thick units of layered sediments from the equator (e.g. Tanaka and Leonard, 1995; Tanaka, 1997; Malin and Edgett, 2000) to the poles (Herkenhoff and Murray, 1990b; Thomas *et al.*, 1992). Transport has been important in moving material for much of Mars' history and the transport has been global in scale. Deciphering the materials and stratigraphy of the wide variety of sedimentary deposits on Mars is crucial to understanding the geological influences of climate on Mars and the relationship of current processes to past ones.

Transport of sediments in the current climate is presently dominated by Hadley circulation in southern summer (Thomas and Gierasch, 1995; Greeley *et al.*, 1992). Many dune deposits, wind streaks, and other deposits of likely aeolian (or possibly lacustrine) origin occur. A primary goal of any study of such deposits is the search for compositional or morphologic clues to their origin, transport, and deposition. A major question exists as to the origin and depositional environments of older sedimentary rocks. Multi-spectral IR, nighttime temperature, and visible observations will be used to map the morphologies, compositional units, and physical properties in key areas identified from earlier missions.

The polar deposits, thought to be climatically sensitive because of their association with frost deposition, include at least two non-volatile components as indicated by Viking Orbiter color data (Herkenhoff and Murray, 1990a; Thomas and Weitz, 1989). The layered deposits appear to reflect cycles of deposition and erosion on a variety of time scales, and their overall extent has been reduced from previously larger deposits (Thomas *et al.*, 1992). Coverage of selected areas of the layered deposits, chasmata, deposits marginal to the layered deposits, and polar dunes with up to 14 visible and IR filters will provide discrimination of surface units beyond what is possible using Viking images, TES 5-km resolution spectra, and MOC panchromatic images. The additional compositional and particle texture discrimination available in the infrared and color image will improve the ability to discriminate and interpret spectrally distinct surface units. A spatial resolution of 18 meters per pixel permits local morphology to be associated with compositional units. This is particularly useful in relating materials exposed on dunes and wind streaks to current wind regimes, and relating morphologies to source areas, and in examining small-scale features thought to be related to seasonal frost cycling (Thomas *et al.*, 2000).

2.2.3. *Craters as Stratigraphic Probes*

In addition to layered materials revealed by erosion in crater interior deposits, canyon walls, along faults, and in the polar deposits, additional stratigraphic information can be gleaned from morphological and compositional studies of impact craters. The interior walls of such craters often reveal the layering of the materials that were impacted, while the ejecta blankets show radial compositional gradations reflecting the inversion of stratigraphy in the ejecta (e.g. Shoemaker, 1963). The visible and IR observations can be combined with spatial studies (i.e., observations of nearby craters of similar size) to trace subsurface composition and structure.

2.3. TEMPERATURE ANOMALIES

The Mars Odyssey orbit is ideally suited for the detection of pre-dawn temperature anomalies not associated with solar heating. Multi-spectral temperature maps with a noise-equivalent delta temperature (NEΔT) of 1 K will be produced for the entire planet. These data will allow identification of sites of active hydrothermal systems and potential near-surface igneous activity using the spatial distribution of temperature differences (e.g., along linear zones) to distinguish them from physical properties such as rock abundance. The discovery of active systems would radically alter our current models of the near- surface environment, and we will undertake an exploration for temperature anomalies using pre-dawn temperature measurements.

Regions to be mapped for thermal anomalies will initially focus on young volcanic sites, where mobilization of ground ice would result from intrusive or extrusive volcanic activity. Although it may be unlikely that thermal anomalies are present, their detection would so heavily influence future sampling strategies that a search for them is of high priority.

2.4. POLAR PROCESSES

The two volatiles that are key to understanding Mars climate and evolution are CO_2, as the major agent for transport and storage of energy, and H_2O, as an important geomorphic agent. Models of their physical properties, their sublimation, and their seasonal behavior have been developed which agree with the general features of existing observations, and allow some extrapolation into the past (e.g. Davies *et al.*, 1977; Kieffer, 1979, 1990; James and North, 1982; Haberle and Jakosky, 1990; Kieffer and Zent, 1992; Kieffer *et al.*, 2000). At present, however, our understanding of the martian poles is sufficiently incomplete to prevent detailed modeling and predictions of volatile behavior in either time or space. Several geophysical processes remain to be fully understood, such as the origin and seasonal behavior of unusual 'Cryptic' regions (Kieffer *et al.*, 2000), the complex physical nature of the polar ices (Kieffer *et al.*, 2000; Titus *et al.*, 2001), the incorporation of dust and water ice into the seasonal polar caps, the influence of suspended aerosols on the radiation balance over the winter cap (e.g. Paige and Ingersoll, 1985) and the processes which cause the major polar outliers.

The THEMIS investigation will address a variety of polar cap processes using moderate spectral and spatial resolution observations. Specific questions include: (1) what is the spatial and seasonal variation of polar condensates? (2) what produces the observed differences in the physical state of the surface condensates (e.g. the south polar Cryptic region (Kieffer *et al.*, 2000)); (3) what is the energy balance in the cap-atmosphere system? (4) what is the role of dust in influencing the surface frost accumulation and sublimation? and (5) what is the abundance and variability of atmospheric dust over the polar caps?

2.5. ATMOSPHERIC TEMPERATURE AND OPACITY

A range of atmospheric properties will be addressed by the THEMIS experiment, including: (1) the abundance and distribution of atmospheric dust, (2) the distribution and condensate abundance of H_2O clouds; and (3) the atmospheric temperature, which will be used to study atmospheric dynamics. THEMIS extends the systematic monitoring of temperature, dust, and water-ice clouds in the martian atmosphere begun by the Mariner 9 IRIS (Conrath *et al.*, 1973), Viking IRTM (Kieffer *et al.*, 1977; Martin, 1986), and TES investigations (Conrath *et al.*, 2000; Smith *et al.*, 2001a, 2002; Pearl *et al.*, 2001). The THEMIS instrument is greatly limited in its ability to map the vertical distribution of temperature and to observe water vapor relative to the TES. THEMIS does, however, provide observations similar to IRTM for the mapping of temperature (Kieffer *et al.*, 1977), ice (Christensen and Zurek, 1984; Tamppari *et al.*, 2000), and dust (Martin, 1986) and will extend the detailed year-to-year climate record begun by the MGS TES.

3. Thermal Infrared Remote Sensing

3.1. THERMAL INFRARED EMISSION SPECTROSCOPY

3.1.1. *Overview*
Surface composition can be determined remotely by two basic types of optical spectroscopy: (1) moderate-energy electronic spectroscopy, which samples the electron cloud surrounding the nucleus; and (2) low-energy vibrational spectroscopy, which samples bound atoms within a crystal structure.

Electronic transitions occur when a bound electron absorbs an incoming photon and transitions to a higher energy state. These energy states are quantized and vary with atomic composition and crystal structure, thus providing diagnostic information about the elemental and crystal structure. The transition metals, of which iron is common, have electronic transitions that occur in the visible and infrared (e.g. Burns, 1993), and visible and near-IR observations provide an excellent means for studying Fe-bearing minerals (McCord *et al.*, 1982; Singer, 1982; Morris *et al.*, 1990; Clark *et al.*, 1990; Bell *et al.*, 1990; Mustard *et al.*, 1993; Mustard and Sunshine, 1995).

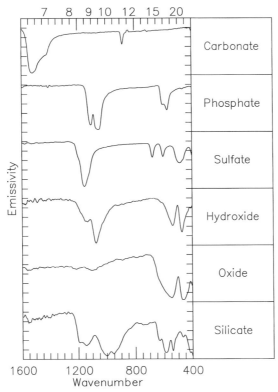

Figure 3. Thermal emission spectra of major mineral classes. These representative spectra show the significant differences in the fundamental vibrational bands between different mineral groups. Individual spectra have been offset and scaled for clarity. The band depths in these coarse particulate mineral samples have a band depth (emissivity minima) relative to the nearby local emissivity maxima of 0.2 to 0.6.

Vibrational spectroscopy is based on the principle that vibrational motions occur within a crystal lattice at frequencies that are directly related to the crystal structure and elemental composition (i.e. mineralogy) (e.g. Wilson *et al.*, 1955; Farmer, 1974). The fundamental frequencies of geologic materials typically correspond to wavelengths greater than ~ 5 μm, and provide a diagnostic tool for identifying virtually all minerals.

An extensive suite of studies over the past 35 years has demonstrated the utility of vibrational spectroscopy for the quantitative determination of mineralogy and petrology (e.g. Lyon, 1962; Lazerev, 1972; Vincent and Thompson, 1972; Farmer, 1974; Hunt and Salisbury, 1976; Salisbury *et al.*, 1987a, b, 1991; Salisbury and Walter, 1989; Bartholomew *et al.*, 1989; Salisbury, 1993; Christensen and Harrison, 1993; Lane and Christensen, 1997; Feely and Christensen, 1999; Christensen *et al.*, 2000a; Hamilton, 2000; Hamilton and Christensen, 2000; Wyatt *et al.*, 2001; Hamilton *et al.*, 2001). The fundamental vibrations within different anion groups, such as CO_3, SO_4, PO_4, and SiO_4, produce unique, well separated spectral bands

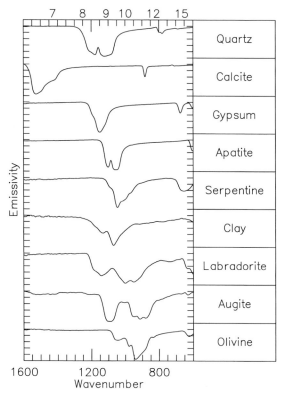

Figure 4. Thermal emission spectra of key minerals. Representative hydrothermal and aqueous minerals are shown, along with examples of the major volcanic-rock forming minerals. Individual spectra have been offset and scaled for clarity.

that allow carbonates, sulfates, phosphates, silicates, and hydroxides to be readily identified (Figure 3). Additional stretching and bending modes involving major cations, such as Mg, Fe, Ca, and Na, allow further mineral identification, such as the excellent discriminability of minerals within the silicate and carbonate groups (Figure 4). Significant progress also has been made in the development of quantitative models to predict and interpret the vibrational spectra produced by emission of energy from complex, natural surfaces (e.g. Conel, 1969; Henderson et al., 1992; Hapke, 1993; Salisbury et al., 1994; Moersch and Christensen, 1995; Wald and Salisbury, 1995; Mustard and Hays, 1997).

The fundamental vibrations of geologic materials typically occur between ~ 6 and 100 μm. In addition to these modes, overtone and combination vibrations, such as the 2.35 μm ($3\nu_3$) and 2.55 μm ($\nu_1 + 2\nu_3$) combination tones in carbonates (Gaffey, 1984) and the 4.5 μm ($2\nu_3$) overtone in sulfate (Blaney and McCord, 1995), also occur. These vibrations typically occur between ~ 2 and ~ 6 μm (Roush et al., 1993). While they also contain important diagnostic information, these modes are typically much less populated than the fundamental vibrations (Wilson et al.,

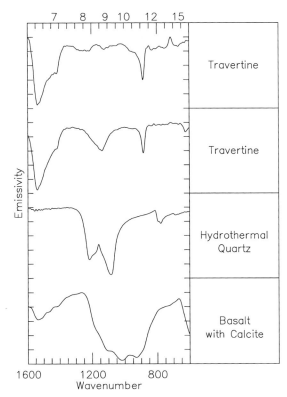

Figure 5. Thermal emission spectra of hydrothermal rocks. The travertine and hydrothermal quartz samples are from a hot-spring system within a volcanic environment. The hydrothermally altered basalt sample contains small (< 1 mm) calcite-bearing vesicles, demonstrating that < 5% carbonate can be detected in volcanic rocks. Individual spectra have been offset and scaled for clarity.

1955). As a result, the overtone and combination band absorptions in the 2–6 μm region tend to be relatively weak compared to the fundamental absorptions in the 6 – 100 μm region.

3.1.2. *Mineral Groups*
3.1.2.1. *Hydrothermal Minerals.* Thermal springs will be excellent candidates for exploration (Walter and Des Marais, 1993) and produce characteristic mineralization that is dominated by microcrystalline quartz (chert, chalcedony, opal, etc.) and carbonates. The silica minerals have a major 8–10 μm absorption (e.g. Hunt and Salisbury, 1970). Carbonates precipitate in thermal spring environments, and are the key constituents of the martian meteorite samples examined by McKay et al. (1996). The fundamental C-O absorption occurs near 6.7 μm (e.g. Farmer, 1974; Nash and Salisbury, 1991; Lane and Christensen, 1997) in a region that is distinct from other mineral classes (Figures 3 and 4), thereby greatly facilitating carbonate identification.

Figure 5 shows laboratory thermal emission spectra of travertine and hydrothermal silica samples collected from the Castle Hot Springs Volcanic Field of central Arizona. The distinctive spectral character of both types of hydrothermal deposits in the thermal infrared are apparent. Travertine samples are characterized by the broad absorption features typical of carbonates (calcite). The hydrothermal silica spectrum clearly exhibits the major absorption features typical of quartz. The basalt sample is an excellent analog for hydrothermal alteration on Mars. It contains small (< 1 mm) calcite-bearing vesicles and veins, similar to those found in the SNC sample ALH84001 (McKay et al., 1996). The spectrum of this rock demonstrates that a small amount of carbonate ($< 5\%$) can be detected in volcanic rocks using thermal-IR spectra.

3.1.2.2. *Evaporite Minerals.* This broad class of minerals includes the following important groups: carbonates, sulfates, chlorides, and phosphates that are precipitated by the evaporation of marine or nonmarine waters. As such, they are the most obvious and direct mineralogical evidence for standing water. In the search for candidate sites for sample return, locating evaporite minerals is a high priority. The abundance of any one of these minerals in an evaporite basin is a function of the dissolved chemical constituents contained in the water as well as the history of the basin inundation/denudation. Therefore, the identification and quantification of the different evaporite minerals can yield information about the environment in which they were produced. Thermal-infrared spectra provide distinguishing characteristics for the different groups. Sulfates (gypsum) and phosphates (apatite) have deep, well-defined features in the 8.3 to 10 μm region that vary with position based on composition (Figure 4).

3.1.2.3. *Hydrous Silicates.* Minerals that incorporate hydroxyl (OH)- anions into their structure give clues about the availability of water during their formation. The majority of such minerals occur in the silicate class and most of these are in the phyllosilicate group. Within the phyllosilicates, the clay, mica, serpentine, and chlorite groups are all important. Serpentine minerals form through the activity of H_2O and ultramafic igneous rocks, so they may provide additional evidence of hydrothermal activity on Mars. Though all the hydrous silicates have the hydroxyl anion as their common trait, they range widely in their mode of occurrence. Some form as primary constituents of igneous rocks, giving clues about the magmatic conditions under which the rock was formed. Most hydrous silicates crystallize as secondary products of metamorphism and hydrothermal alteration and their composition provides insight into the pressure and temperature where they formed. Thus, the hydrous silicates serve as excellent pathfinder minerals for hydrous activity. All have characteristic mid-IR features (e.g. Figure 4) due to fundamental bending modes of $(OH)^-$ attached to various metal ions, such as an AL-O-H bending mode near 11 μm in kaolinite clay (e.g. Farmer, 1974; Van der Marel and Beeutelspacher, 1976).

3.1.2.4. *Igneous Silicates.* The primary silicate minerals associated with igneous rocks are the most abundant mineral class found on Mars (Christensen *et al.*, 2000b, 2001a; Bandfield *et al.*, 2000a; Bandfield, 2002). The majority of martian rocks likely will vary by only relatively subtle differences in bulk mineralogy, represented by the common rock forming minerals. An ability to distinguish and quantify olivines, pyroxenes, and feldspars is crucial to describing the geological character of the planet. Without this overview, the locations of hydrous activity are without context. All silicates have Si-O stretching modes between 8 and 12 μm that vary in position with mineral structure (e.g. Figure 4). This absorption shifts to higher frequency (shorter wavelength) as bond strength increases for isolated, chain, sheet, and framework tetrahedron structure. These shifts allow for detailed identification of the igneous silicates, including variations within the solid solution series.

3.1.3. *Quantitative Analysis of IR Spectra*
A key strength of mid-infrared spectroscopy for quantitative mineral mapping lies in the fact that mid-infrared spectra of mixtures are linear combinations of the individual components (Thomson and Salisbury, 1993; Ramsey, 1996; Feely and Christensen, 1999; Hamilton and Christensen, 2000). The mid-IR fundamental vibration bands have very high absorption coefficients and therefore much of the emitted energy only interacts with a single grain. When absorption coefficients are low, as is the case for overtone/combination bands, the energy is transmitted through numerous grains and the spectra become complex, non-linear combinations of the spectral properties of the mixture. The linear nature of the thermal spectral emission of mineral mixtures has been demonstrated experimentally in particulates for mixtures of up to five components (Thomson and Salisbury, 1993; Ramsey, 1996). In these experiments the mineral abundance could be quantitatively retrieved using linear deconvolution techniques to within 5% on average. The linear mixing of mineral components in rock spectra has also been confirmed (Feely and Christensen, 1999; Hamilton and Christensen, 2000; Wyatt *et al.*, 2001; Hamilton *et al.*, 2001), with retrieved mineral abundances that are accurate to 5–10% in laboratory spectra.

The successful determination of mineral composition and abundance is illustrated in Figure 6. Mineral composition and abundance were determined both spectroscopically and using traditional thin-section techniques for a suite of 96 igneous and metamorphic rocks (Feely and Christensen, 1999). The rocks were used in their original condition; no sample cutting, polishing, or powdering was performed, and weathered surfaces were observed where available to best simulate remote observations. Comparison of the mineral abundances determined spectroscopically with the petrographically estimated modes for each sample gave an excellent agreement using high-resolution data (Figure 6a). The spectroscopically determined compositions matched the petrologic results to within 8–14% for quartz, carbonates, feldspar, pyroxene, hornblende, micas, olivine, and garnets. These val-

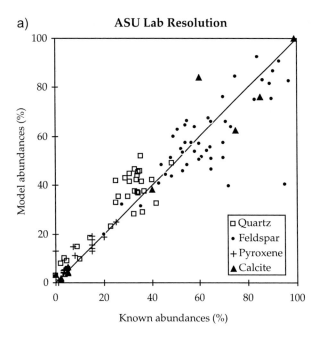
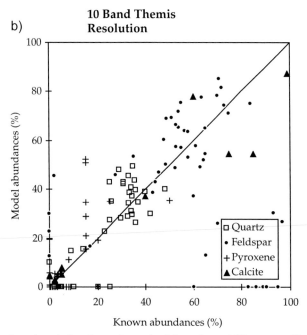

Figure 6. Quantitative mineral abundance determined from thermal-IR spectra. 'Known' abundances are derived from optical thin section measurements; 'model' abundances are derived from linear deconvolution of infrared spectral data (Feely and Christensen, 1999). The solid line corresponds to a perfect match between the two methods. (a) Mineral abundance derived from high spectral resolution (TES-like) data. (b) Mineral abundance derived from 10-band (THEMIS-like) data.

ues are comparable to the 5–15% errors typically quoted for traditional thin section estimates.

The mineral abundances derived using 10-band THEMIS-like resolution for the same hand sample rock suite are shown in Figure 6b. These results demonstrate that mineral abundance can be determined to within 15% using only 10 thermal-IR spectral bands on naturally-occurring rock surfaces. Similar results have been achieved for remotely-sensed aircraft observations on Earth through an atmosphere on a planetary surface (Ramsey *et al.*, 1999). Laboratory studies have also shown that a signal-to-noise ratio (SNR) of 35-100, corresponding to a noise-equivalent delta emissivity (NE$\Delta\varepsilon$) of 0.03 to 0.01, is sufficient to identify minerals using thermal-IR data (Feely and Christensen, 1999).

3.1.4. *Environmental Effects*

Variations in particle size and porosity produce variations in the spectra of materials at all wavelengths. Numerous quantitative models have been developed to investigate these effects (Vincent and Hunt, 1968; Hunt and Vincent, 1968; Conel, 1969; Hunt and Logan, 1972; Hapke, 1981, 1993; Salisbury and Eastes, 1985; Salisbury and Wald, 1992; Salisbury *et al.*, 1994; Moersch and Christensen, 1995; Wald and Salisbury, 1995; Mustard and Hays, 1997) and have demonstrated the importance of specular reflectance and scattering. Two basic behaviors are observed with decreasing grain size: (1) strong bands (high absorption) tend to get shallower; and (2) weak bands (low absorption) increase in contrast, but appear as emission maxima and reflectance minima (Vincent and Hunt, 1968).

Dust coatings and weathering rinds present a potential problem for any optical remotely-sensed measurements of Mars. However, the thickness of material through which sub-surface energy can escape increases linearly with wavelength. Thermal IR spectral measurements through coatings have been studied using mechanically deposited dust (Ramsey and Christensen, 1992; Johnson *et al.*, in press) and terrestrial desert varnish (Christensen and Harrison, 1993) as analogs to martian rock coatings. These results have shown that thermal-IR spectral observations can penetrate relatively thick (mean thickness up to ~ 40–50 μm) layers of these materials to reveal the composition of the underlying rock.

Atmospheric dust is also an issue in the remote sensing of the martian surface (Bandfield *et al.*, 2000b; Smith *et al.*, 2000; Christensen *et al.*, 2000b). However, scattering and absorption by fine-grained (< 5 μm) dust suspended in the atmosphere at typical opacities of < 0.2 (Smith *et al.*, 2001b, c) produces a linear contribution to the infrared spectrum (Smith *et al.*, 2000), and methods have been developed to quantitatively remove this contribution from THEMIS multi-spectral IR images (Smith *et al.*, 2000; Bandfield, 2002).

3.2. THERMAL INFRARED MULTI-SPECTRAL IMAGING

It has been established that minerals can be accurately identified in mixtures given high spectral resolution data, such as are available in a laboratory or from the MGS TES instrument (Ramsey and Christensen, 1998; Feely and Christensen, 1999; Hamilton, 1999; Christensen et al., 2000b, c; Bandfield et al., 2000a; Bandfield, 2002). The key question regarding the use of multi-spectral imaging is to determine what spectral resolution is sufficient to determine the presence and abundance of aqueous minerals. Thermal-infrared observations acquired in six spectral bands from 8 to 12 μm have proven extremely powerful for geologic mapping on Earth, and have clearly demonstrated that subtle differences in rock composition can be mapped (Gillespie et al., 1984; Kahle et al., 1980; Crisp et al., 1990; Hook et al., 1994; Edgett and Christensen, 1995; Kahle et al., 1993; Ramsey, 1996).

The spectra of the key minerals shown in Figures 4 and 5 at TES spectral resolution are reproduced in Figure 7 at THEMIS resolution by convolving the laboratory spectra with the THEMIS filter response functions (see Section 4.2). Carbonates are identified using bands in calcite absorption (6.6 μm) and the continuum (7.5 μm) region. Silica (quartz) is also readily separated at this resolution from the other silicates, which have lower frequency absorptions. Key evaporite minerals, gypsum and apatite, also have well defined bands. Hydrated minerals (serpentine and clay) have unique absorption band shapes and positions, as do the common volcanic minerals, labradorite and augite, which have two very well defined absorptions in this spectral range.

Figure 8 provides an example of the rock composition mapping that can be achieved using broad-band multi-spectral thermal-IR data. This image was acquired using the 6-band Thermal Infrared Multispectral Scanner (TIMS) imager operating from 8–14 μm (Palluconi and Meeks, 1985) over the Granite Wash Mountains, Arizona. Three bands centered at ∼ 8, 9, and 10 μm are displayed in blue, green, and red, respectively. Different rock types are readily identified and their spatial distribution mapped using these thermal-IR multispectral data, as confirmed by detailed geologic mapping.

This image illustrates an important attribute of thermal-IR spectroscopy, where the spectra are sensitive to the major rock-forming minerals, rather than minor impurities, stains, and coatings. The 3-band data clearly discriminate carbonate (green), quartz-rich (red), and clay-rich (light blue), sedimentary rocks (purple/yellow), as well as basaltic (blue) and andesitic volcanic rocks (pink/purple). Soil surfaces of differing ages appear red to orange due to differing weathering processes that produce and remove clays and concentrate more resistant quartz-rich clasts over time.

The THEMIS 8-wavelength spectra will allow mineral identification and a quantitative determination of mineral abundance. The power of this combination of imagery and spectroscopy can be seen using spectra of rocks from the Granite Wash region convolved to the THEMIS surface-sensing bandpasses (Figure 9). Spectra

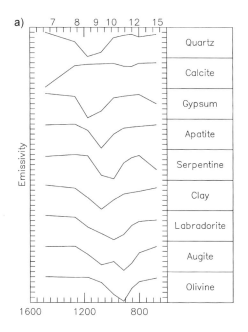
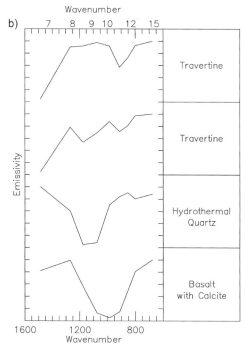

Figure 7. The emission spectra of key minerals at THEMIS spectral resolution. (a) The minerals from Figure 4 are shown at THEMIS resolution. The emissivity band depths vary from 0.2 to 0.4, well above the THEMIS noise level. (b) The hydrothermal rocks from Figure 5 are shown at THEMIS resolution.

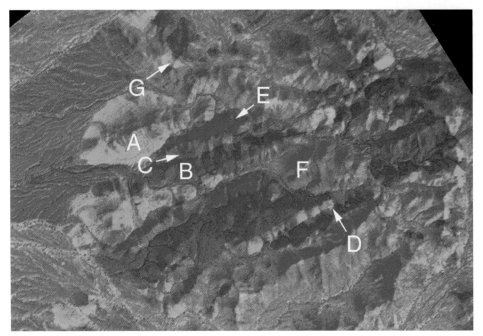

Figure 8. Infrared multispectral composition mapping. This three-band terrestrial IR image illustrates the rock and mineral discrimination using only three wavelength bands. Red surfaces are silica rich rocks and sediments, green units are carbonates, purple units are quartz-rich volcanic sediments, and blue units are mafic volcanics and metavolcanics. Also shown are the locations of the rock samples whose spectra are given in Figure 9.

were measured from unprepared, weathered and coated surfaces to duplicate remote measurements. The samples include fine-grained rocks with microcrystalline textures including those of hydrothermal origin, silica-rich rocks, quartzite, carbonate (limestone), a gypsum-rich soil, a hydrothermally altered basalt, and an ultra-mafic amphibolite.

The spectra of this suite of rocks at THEMIS resolution are unique and their mineral composition can be determined (Figure 9). Carbonates and quartz can be easily identified, as can gypsum, which appears bright yellow in the limited regions where it occurs in Figure 8. The sample of the microcrystalline quartz rock containing calcite veins has been hydrothermally altered and both carbonate and quartz are readily detected. The mafic rocks typical of expected martian volcanic rocks are distinct and easily separated from the aqueous samples.

The rocks in the Granite Wash image are heavily coated with desert varnish. These coatings are optically thick at visible wavelengths, with white sandstone appearing dark brown to black in the field.. However, as the Granite Wash image data demonstrate, the mid-infrared spectral signatures of the underlying rocks are identified through these coatings that are estimated to be up to 50 μm thick.

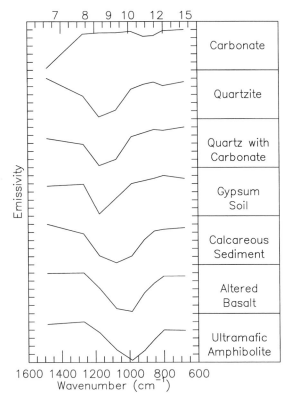

Figure 9. THEMIS simulated 8-band spectra of rock samples collected from the field sites shown in Figure 8. These spectra demonstrate the spectral discrimination possible with only 8 surface-sensing IR bands.

4. Instrumentation

4.1. MEASUREMENT REQUIREMENTS

THEMIS has two primary mission objectives: (1) to identify minerals and compositional units at 100-m spatial scales; and (2) to resolve surface features at scales significantly less than 100 m. The mineral mapping objective has three measurement requirements: (1) radiometric precision and accuracy necessary to resolve the expected band depths for minerals present at 10% abundance; (2) spectral resolution sufficient to identify key minerals; and (3) a spatial resolution sufficient to isolate small mineral deposits ($\sim$ 100 m). The morphology objective results in a measurement requirement of an instantaneous field of view (IFOV) that is a factor of 3–5 times less than 100 m.

The radiometric requirements are best stated relative to the band depth of minerals likely to be found on the martian surface weighted by their abundance. The band depth is defined here as the minimum emissivity in an absorption band relative to an emissivity of 1.0. The emissivity minima of particulates of the key quartz and carbonate minerals are ∼ 0.5 (band depth 0.5) in the 6–14 μm range; sulfates, phosphates, and clays are ∼ 0.6 (band depth 0.4) ; typical volcanic minerals pyroxene, plagioclase, and olivine range from 0.7–0.8 (band depth 0.3–0.2). A surface containing carbonate grains at 10% abundance with an emissivity of 0.5 at 6.6 μm, mixed with 90% silicate with an emissivity of 1.0 would produce an absorption feature with an emissivity of 0.95. In order to observe this absorption feature, a noise equivalent delta emissivity (NE$\Delta\varepsilon$ of ∼ 0.02 would be required. (Note that the signal-to-noise ratio (SNR) is the reciprocal of NE$\Delta\varepsilon$). The imaging nature of THEMIS will permit spatial aggregation of pixels and a spatial context to be developed, again increasing the acceptable noise level in each pixel (the detection of interesting sites will not depend on the occurrence of 10% carbonate in a single pixel). Near 10 μm where a number of minerals have absorption features, it is desired to have a higher NE$\Delta\varepsilon$ to allow mineral discrimination. Thus, the minimum NE$\Delta\varepsilon$ requirement for THEMIS is 0.02 (SNR = 50) in the 6.6 μm band and 0.007 (SNR = 143) in the 8–12.5 μm bands for each pixel.

The average local time of the THEMIS multi-spectral observations is 4:30 PM. The surface temperature follows the sub-solar point, but remains $\geq$ 245 K at this time over a wide latitude range for all thermal inertia surfaces (Kieffer *et al.*, 1977). For example, at 4:30 PM in southern hemisphere summer, the temperatures are between 245 and 270 from 45° S to 30° N, and in northern hemisphere summer the temperatures range from 235 to 250 K between 0° and 45° N. These latitude ranges encompass expected future rover/lander allowable ranges, and include 70% of the total surface area of Mars. A conservative requirement of a surface temperature of 245 K has been selected to set the instrument performance requirements.

The NE$\Delta\varepsilon$ requirement translates to a NEΔT requirement of 0.9 K at 7 μm and 0.5 K at 10 μm for a surface temperature of 245 K. The identification of nighttime temperature anomalies can be achieved with a NEΔT of 1 K at a typical nighttime surface temperature of 180 K.

4.2. GENERAL DESCRIPTION

The design of the THEMIS is intentionally conservative. We have adopted a multi-spectral, rather than hyperspectral, approach that is sufficient to quantitatively determine mineralogy and allows global coverage within the available data volume. The THEMIS flight instrument, shown in Figure 10, consists of infrared and visible multi-spectral imagers that share the optics and housing but have independent power and data interfaces to the spacecraft to provide system redundancy. The details of the instrument design are given in Table 1. The telescope is a three-mirror anastigmat with a 12-cm effective aperture and a speed of f/1.6. A calibration flag, the only moving part in the instrument, provides thermal calibration and is used to protect the detectors from direct illumination from the Sun. The electronics provide digital data collection and processing as well as the instrument control and data

THERMAL EMISSION IMAGING

Figure 10. THEMIS flight instrument.

TABLE I

Themis instrument design summary

Quantities to be measured:	Emitted radiance in 10 ~1-μm bands (9 different wavelengths) centered between 6.8 and 14.9 μm at 100 m per pixel spatial resolution
	Solar reflected energy in 5 ~50 nm bands centered from 0.42 to 0.86 μm at 18 m per pixel spatial resolution
Detectors:	Multi-spectral IR imager: 320 × 240 element uncooled microbolometer array
	Visible imager: 1,024 × 1,024 element silicon array
Expected performance:	NE$\Delta\varepsilon$ @ 245 K & 10 μm = 0.005.
	NEΔT @ 245 K & 10 μm = 0.5 K.
	NEΔT @ 180 K = 1 K.
	SNR > 100 in visible imager at 0.25 albedo, 60° solar incidence angle
Optical/Mechanical Design:	12-cm effective aperture f/1.6 telescope to view nadir shared by multi-spectral IR and visible imagers. Internal calibration flag provides calibration and Sun avoidance protection functions. IR spectrometer is 10-strip filter pushbroom design with time delay integration. Visible multi-spectral imager is 5-strip filter frame-scan design.
Fields of View:	IR imager has a 4.6° (80 mrad) cross-track by 3.5° (60 mrad) down-track FOV with a 0.25 mrad (100 m) IFOV at nadir.
	Visible imager has a 2.66° (46.4 mrad) cross-track by 2.64° (46.1 mrad) down-track FOV with 0.045 mrad (18 m) IFOV in 1024 × 1024 pixels at nadir. The two imagers are spatially bore-sighted.
In-Flight Calibration:	Periodic views of an internal calibration flag.
Thermal Requirements:	Operating range −30 °C to +30 °C. Non-operating range −30 °C to +45 °C. No detector cooling required.
Digital Data	8-bit delta radiance in IR imager
	8-bit radiance in visible imager
Data rate:	IR imager has instantaneous internal rate of 1.17 Mbits/sec. Data rate to spacecraft after real-time compression is 0.6 Mbits/sec.
	Visible imager has instantaneous internal rate of up to 6.2 Mbits/sec. Data rate to spacecraft is < 1.0 Mbits/sec. 4 Mbyte RAM internal data buffer for data processing and buffering for delayed output to spacecraft.
On-board Data Processing:	IR imager: Gain and offset; time delay integration, and data compression in electronics; data formatting using spacecraft computer.
	Visible imager: Lossless and predictive compression in firmware; selective readout and pixel summing on spacecraft.
Solar Protection:	Provided by calibration flag in stowed position.
Mass:	11.2 kg
Size:	29 cm by 37 cm by 55 cm
Power:	14 W orbital average

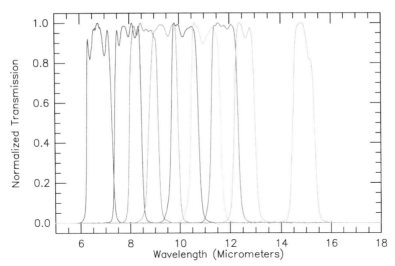

Figure 11. THEMIS IR imager spectral bandpasses. Data were collected at SBRS prior to instrument delivery to the spacecraft.

interface to the spacecraft. The instrument weighs 11.2 kg, is 29 cm by 37 cm by 55 cm in size, and consumes an orbital average power of 14 W.

A major feature of the instrument is the use of an uncooled microbolometer array operated at ambient temperature, which substantially reduced the complexity of instrument fabrication, testing, spacecraft interfaces, and mission operations. This array has 320 cross-track pixels and 240 down-track pixels with an IFOV of 100 m and an image width of $\sim$ 32 km. A small thermal electric (TE) cooler is used to stabilize the IR focal plane temperature to ± 0.001 K. The IR imager has ten stripe filters that produce ten $\sim$ 1-μm wide bands at nine separate wavelengths from 6.78 to 14.88 μm (Table 2; Figure 11). Two filters (Bands 1 and 2) cover the same spectral region centered at 6.78 μ to improve the detection of carbonate by improving the signal to noise in this spectral region. The nine IR wavelengths include eight surface-sensing wavelengths (Bands 1–9) and one atmospheric wavelength (Band 10).

The visible imager is a derivative of the Malin Space Science Systems (MSSS) Mars Polar Lander Mars Decent Imager (MARDI), with a 5-filter subset of the MSSS Mars Color Imager (MARCI) developed for the Mars Climate Orbiter. It has 1,024 cross-track pixels with an 18-m IFOV covering a 18.4-km swath boresighted with the IR imager through a beamsplitter. The visible imager has five stripe filters centered from 0.425 to 0.86 μm (Table 3; Figure 12).

4.3. OPTICAL DESIGN

In order to integrate the visible and IR bands into a single telescope, a fast, wide field-of-view reflective telescope has been used. The 3.5° (down-track) × 4.6°

TABLE II
THEMIS infrared band characteristics

Band	Center Wavelength (μm)	Half Power Point – Short Wavelength (μm)	Half Power Point – Long Wavelength (μm)	Band Width (Full Width Half Max) (μm)	SNR
1	6.78	6.27	7.28	1.01	33
2	6.78	6.27	7.28	1.01	34
3	7.93	7.38	8.47	1.09	104
4	8.56	7.98	9.14	1.16	163
5	9.35	8.75	9.95	1.20	186
6	10.21	9.66	10.76	1.10	179
7	11.04	10.45	11.64	1.19	193
8	11.79	11.26	12.33	1.07	171
9	12.57	12.17	12.98	0.81	132
10	14.88	14.45	15.32	0.87	128

(cross-track) field of view is achieved with a 3 mirror f/1.6 anastigmat telescope with an effective aperture of 12 cm and a 20-cm effective focal length. The design allows for excellent baffling to minimize scattered light. It is based on a diamond-

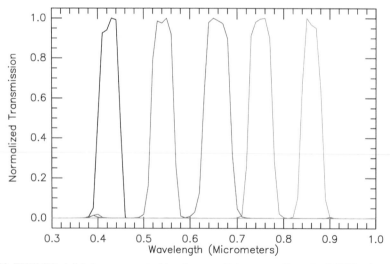

Figure 12. THEMIS visible imager spectral bandpasses. Data were collected at MSSS prior to visible sensor delivery to SBRS.

TABLE III
THEMIS visible band characteristics

Band	Center Wavelength (μm)	Half Power Point – Short Wavelength (μm)	Half Power Point – Long Wavelength (μm)	Bandwidth (Full Width Half Power) (μm)
1	0.425	0.400	0.449	0.049
2	0.540	0.515	0.566	0.051
3	0.654	0.628	0.686	0.053
4	0.749	0.723	0.776	0.053
5	0.860	0.837	0.882	0.045

turned bolt-together approach to telescope design, fabrication, alignment and test. A telescope ray trace is shown in Figure 13.

The system was optimized to match the high signal performance required for the IR imager and the spatial resolution needed for the visible camera. The 9-μm pitch of the visible array maps to a ground sample distance (GSD) of 18 meters with an MTF of approximately 0.1 at Nyquist. Similarly, the 50 μm pitch of the IR focal plane array maps to a GSD of 100 M.

A major aspect of THEMIS is the use of diamond-turned 'bolt-together' telescope technology. The telescope development used design and analysis techniques that allowed complex off-axis designs to be fully modeled. The manufacture utilized high-precision machining capabilities that allowed the entire optical stage to be machined and assembled without manual optical component adjustments, and achieved diffraction-limited performance in both the visible and infrared. Specified mounting surfaces were machined with extremely tight tolerances ($\pm$ 0.0002 in). The optical surfaces were machined directly from high order aspheric equations. The telescope was manufactured with aluminum to reduce cost and is significantly light-weighted. Nickel plating and automated post polishing were used to keep the surface scatter to levels unobtainable with conventional diamond turning techniques.

4.4. FOCAL PLANE ASSEMBLIES

THEMIS uses uncooled microbolometer detector arrays that have been recently declassified and were produced commercially by the Raytheon Santa Barbara Research Center (SBRC) under license from Honeywell, Inc. The THEMIS IR imager design is based on a Raytheon hand-held imager developed for rugged military use, thus significantly reducing the development cost of the IR focal plane assembly. The microbolometer array contains 320 pixels cross track by 240 pixels along

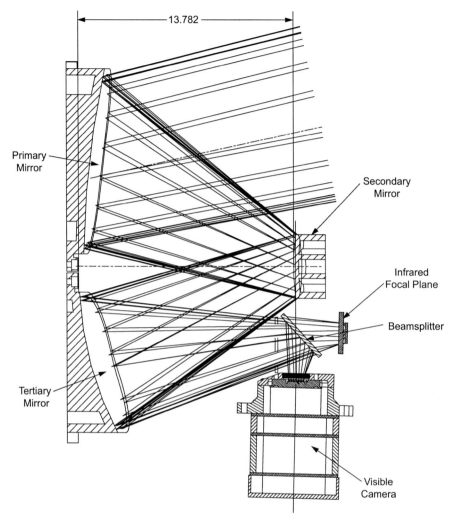

Figure 13. The THEMIS optical schematic and raytrace.

track, with a square 50 micron pixel pitch. The microbolometer arrays were grown directly on the surface of Readout Integrated Circuits (ROIC) which are designed by SBRC and utilize custom Digital Signal Processing electronics.

Spectral discrimination in the infrared is achieved with ten narrowband stripe filters. Each filter covers 16 lines in the along track direction with an 8-line 'deadspace' between filters. To maximize manufacturing yield and reduce costs, the stripe filters were fabricated as separate stripe filters butted together on the focal plane as shown in Figure 14. The THEMIS arrays and filters are mounted within a dewar assembly that seals the arrays and allows for simplified ambient testing. The along-track detectors under a common spectral filter are combined by the use of time-delay and integration (TDI) to improve the instrument's signal-to-noise

Figure 14. The THEMIS infrared focal-plane and stripe-filter layout.

performance. The calculated dwell time for a single pixel, at a martian orbit of 400 km and a 100-meter footprint is 29.9 msec, which closely matches the 30 Hz frame rate for the standard microbolometer.

The visible camera supplied by MSSS consists of a small (5.5 × 8.5 × 6.5 cm, < 500 gm) unit incorporating a focal plane assembly with five color filters superimposed on the CCD detector, a timing board, a data acquisition subsystem and a power supply. This sensor utilizes a Kodak KAI-1001 CCD that was flown in the MS'98 MARDI instrument. This detector has 1024 × 1024 9-micrometer pixels (1018 × 1008 photoactive). The visible imager uses a filter plate mounted directly over the area-array detector on the focal plane. On the plate are multiple narrowband filter strips, each covering the entire cross-track width of the detector but only a fraction of the along-track portion of the detector. This sensor uses a 5-filter subset of the 10-filter MS'98 MARCI wide angle camera to acquire multispectral coverage. Band selection is accomplished by selectively reading out only part of the resulting frame for transmission to the spacecraft computer. The imager uses 5 stripes each 192 pixels in along-track extent. The entire detector is read out every 1.3 seconds. The five bands selected are centered near 425, 550, 650, 750, and 860 nm (Figure 12)

4.5. ELECTRONICS DESIGN

Both the visible and infrared cameras utilize commercial off-the-shelf electronics with modifications to accommodate space environmental requirements. Dedicated, miniaturized electronics provide ultra-stable, low-noise clock and bias signals to the focal planes, stabilize IR focal plane temperature within ± 0.001 °C, and perform analog and digital processing of the output signals.

The IR signal change begins at the focal plane that generates an analog output. An initial 8-bit analog DC offset correction occurs on the focal plane, followed by analog-to-digital conversion to 12-bit words, which are then corrected for gain and offsets. This correction is provided by the electronics of the IR camera and consists of a 12-bit fine offset and 8-bit gain and responsivity adjustment, performed in real time on a pixel-by-pixel basis. This process eliminates all of the significant noise elements with the exception of the fundamental random noise term. The output from this processing is an 8-bit word for each pixel.

The spacecraft interface electronics supply final processing of the focal plane data, a data and command interface to the spacecraft, and overall instrument power conditioning. Internal THEMIS data processing in firmware includes 16:1 TDI processing and lossless data compression for the IR bands using a hardware Rice data compression algorithm chip. The bulk of the interface electronics is performed using Actel Field Programmable Gate Arrays (FPGAs), that are packaged using a mixture of conventional and Sealed Chip-On-Board, High-Density Multiple Interconnect technology and chip stack memory. The visible and IR subsystems have independent power supplies, the IR power supply uses off-the-shelf modules and requires only a few discrete components for input filtering to assure electromagnetic compatibility with the rest of the spacecraft. The spacecraft processor performs final data stream formatting for both the IR and the visible data.

The visible sensor requires 7 clock signals: a two-phase vertical clock (V1/V2), a two-phase horizontal clock (H1/H2), a sub-state clear clock (S), a reset clock (R), and a fast- dump clock (F). In addition, the ADC requires a convert clock.

The amplified CCD signal is digitized by an Analog Devices AD1672 12-bit ADC running at its maximum rate of 3 MSPS. For each pixel, both reset and video levels are digitized and then subtracted in the digital domain to perform correlated double sampling (CDS).

The digital electronics are responsible for clock generation, sampling of the CCD signal, conversion of the 12-bit samples to 8-bit encoded pixels, storage of the pixels, and finally readout of the pixels to the spacecraft. The zero reference ('reset') level for each pixel is digitized and stored in a register. The sum of the video plus zero reference ('video') level is then digitized, and an arithmetic subtraction is performed to produce the final result. The CCD output only requires scaling to the ADC range; no analog sampling, delay or differencing is required. The digital signal processor within the visible sensor generates the CCD clocks, reads the reset and video levels from the ADC, performs the correlated double sampling subtraction, reduces the pixel from 12 to 8 bits, applies optional 2 × 2 or 4 × 4 pixel summinglossless (2:1) first-difference Huffman compression, and transmits it digitally over the serial communication interface to the spacecraft CPU.

4.6. Mechanical Design

The THEMIS main frame is composed of aluminum and provides the mounting interface to the spacecraft as well as the telescope assembly, thermal blankets, and thermal control surface. The focal plane assemblies are mounted in the main frame using brackets that provide for the necessary degrees of freedom for alignment to the telescope. The calibration shutter flag is stored against a side wall that will maintain a known temperature of the flag for calibration purposes. Aluminum covers are installed over the electronics circuit cards to provide EMI, RFI, and radiation shielding as required. There is no reliance on the spacecraft for thermal control of THEMIS, other than the application of replacement heater power when the instrument is off. The thermal control plan includes the use of multi-layer insulation blankets and appropriate thermal control surfaces to provide a stable thermal environment and a heatsink for the electronics and the TE temperature controller on the focal plane arrays.

4.7. Performance Characteristics

The THEMIS IR performance was estimated in the design phase using expected performance values for the IR focal plane, filters, optical elements, TDI, and electronics. The following equations define the parameters used to predict the noise equivalent delta emissivity (NE$\Delta\varepsilon$):

$$NE\Delta\varepsilon = \frac{NEP}{P_{sc}^* \sqrt{n_{TDI}}}$$

where

NEP = Noise equivalent power

P_{sc} = scene power incident in the bolometer

n_{TDI} = number of samples in TDI (16)

$$NEP = \frac{V_{nz}}{R}$$

where

V_{nz} = noise voltage in volts

R = bolometer responsivity in volts/mW

NEP for the microbolometer is a constant in this analysis for two reasons: (1) bolometer noise is not a function of photon flux (as it is for photon-limited

PV detectors where the shot noise component of the flux dominates); and (2) the responsivity of the bolometer over wavelength is taken as constant and based on that measured in the 'flat' portion of the spectral curve for the bolometer.

The detector output is based on the Mars scene temperature, instrument parameters, and predicted bolometer performance. The radiance calculation is based on Planck's law, multiplied by the optics solid angle (omega) for an average f1.6 system, the area of the microbolometer, the overall optics transmission, and the spectral response predicted for the microbolometer. The resulting power is then divided into the NEP, and improved by the square root of the number of samples in TDI, to yield the NE$\Delta\varepsilon$ ratio. SNR is the reciprocal of this NE$\Delta\varepsilon$. The scene power incident on the bolometer in mW (Psc) is:

$$P_{sc} = L(T_{sc}, l_{sc})^* A_{det}^* \Delta_1^* T_{opt}^* T_{filt}^* T_{bol}(l_{sc})^* \Omega$$

Where the radiance at instrument aperture (L) in mW/cm^2-μm-sr is given by:

$$L(T_{sc}, l_{sc}) = \frac{c_1}{l_{sc}^5} * \frac{1}{e^{c_2/l_{sc}T_{sc}} - 1} * 1000$$

and

l_{sc} = peak wavelength in μm

T_{sc} = scene temperature in °K

$c_1 = 2\pi hc^2 = 37415$ W-μm/cm^4

$c_2 = ch/k = 14388$ μm $-$° K

A_{det} = detector area in cm^2

Δ_1 = bandwidth of the stripe filters in μm

$T_{opt}^* T_{filt}$ = combined optics and filter transmission

T_{bol} = predicted spectral response of the bolometer

Ω = optics solid angle in sr

The predicted performance for the infrared bands produced NE$\Delta\varepsilon$'s ranging from 0.007 to 0.038 when viewing Mars at surface temperatures of 245K to 270 K.

The measured noise equivalent spectral radiance (NESR) values are given in Figure 15. The measured SNR for a reference surface temperature of 245 K is summarized for each band in Table 3. Most of the variation in SNR between bands is due to the variation in emitted energy for the 245 K reference temperature. As seen from these results, the THEMIS IR imager exceeds the proposed measurement requirements by a factor of two in most bands.

Signal-to-noise ratios for the visible imager were computed for a low albedo (0.25), flat-lying surface viewed at an incidence angle of 67.5° under aphelion conditions. The SNR values for this case vary from 200 to 400.

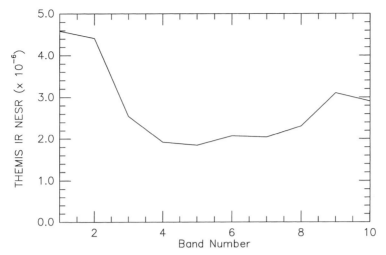

Figure 15. Noise equivalent spectral radiance (NESR) for the 10 THEMIS IR bands. Units are W cm^{-2} μm^{-1} sr^{-1}.

4.8. SOFTWARE

The flight software for the IR imager resides on the spacecraft computer and performs the formatting and data packetization. Instrument commanding is done using discrete spacecraft commands to the THEMIS instrument over an RS-422232 synchronous serial command line. These commands consist of: (1) IR camera on/off/stand-by; (2) visible camera on/off/stand-by; (3) calibration flag shutter control and electronics synchronization; and (4) instrument parameter settings (gain, offset, integration time, etc).

The visible imager software runs on two processors: the main spacecraft CPU and the internal digital signal processor (DSP). The CPU is responsible for instrument operational commands and image post- processing and compression. The DSP is responsible for generating the CCD clocks, emulating the required analog processing and transmitting the data output to the CPU.

The algorithm employed compresses each image line independently by encoding first differences with a single, fixed Huffman table. Selective readout and pixel summing can also be performed by the DSP softwreare. Experience has shown that the volume of data likely to be returned from a spacecraft often evolves during a mission. Implementing data compression in software on the spacecraft computer provides the maximum flexibility for the science and spacecraft team to trade-off data return and buffer space usage. Two compression modes were developed for the VIS: a fast lossless predictive compressor, and a slower lossy discrete cosine transform compressor. Both forms are applied by the spacecraft CPU using THEMIS-provided software. The lossless algorithm compresses each image line independently by encoding pixel first differences with a single, fixed Huffman table.

5. Mission Operations

5.1. OVERVIEW

The Mars Odyssey spacecraft is in a near-polar sun synchronous orbit with an approximately 4:30 PM equator-crossing local time. From this orbit the THEMIS infrared imager will image 100% of the planet both day and night. In addition, THEMIS will be used for selective targeting throughout the mission for regions of high-interest identified from TES, MOC, THEMIS, or other experiments.

Collection of THEMIS IR and visible data is flexible, based on scientific priorities, observing conditions, and data rates. Optimal observing conditions for IR mineralogic mapping occur during mission phases where the local true solar time is the earliest in the afternoon, and when the atmosphere has the lowest dust and water-ice opacity. Optimal conditions for visible imaging occur when the incidence angles are between 60 and 80°. Nighttime images can be acquired during all phases of the mission, but will be emphasized when the daytime IR images are of lower quality. Polar observations will be acquired through the martian year to observed all aspects of autumn cap growth, winter night, and spring retreat.

5.2. IMAGE COLLECTION

THEMIS IR images are acquired at selectable image lengths, in multiples of 256 lines (25.6 km). The image width is 320 pixels (32 km from the nominal mapping orbit) and the length is variable, as given by ((# frames) * 256 lines) - 240. The allowable number of frames varies from 2 to 256, giving minimum and maximum image lengths of 272 and 65,296 lines respectively (27.2 km and 6,530 km from the nominal mapping orbit). The bands returned to the ground are selectable. THEMIS visible images can be acquired simultaneously with IR images, but the spacecraft can only transfer data from one of the two THEMIS imagers at a time. The IR image transfers data as it is being collected, while the visible images are stored within an internal THEMIS buffer for later transfer to the spacecraft computer.

The visible images are acquired in framelets that are 1024 samples crosstrack by 192 lines downtrack in size. Framelets are typically taken at 1-sec intervals, resulting in 26 rows of downtrack overlap at a nominal orbit velocity of 3.0 km/sec. Images can be composed of any selectable combination of bands, image length, and integer pixel summing that can be stored within 3.734 Mbytes THEMIS internal buffer. The size of an image is given by:

$$[(1024*192)*\#framelets*\#bands]/[summing^2] = 3.734 \text{ Mbytes}$$

Thus, for example, with no summing either a single-band, 19-framelet (~ 60 km) image or a 5-band 3-framelet (8.5 km) image can be collected.

THEMIS images are calibrated using periodic views of the internal calibration flag together with an instrument response function determined pre-launch. The flag is closed, imaged, and reopened at selectable intervals throughout each orbit. This process produces gores in the surface observation lasting approximately 50 seconds for each calibration. Calibration data typically are acquired at the beginning and end of each image, but can be spaced to meet the calibration accuracy requirements, while minimizing the loss of surface observations.

5.3. Data allocation

THEMIS data collection will be distributed between the mineralogic, temperature, and morphologic science objectives in both targeted and global mapping modes. A baseline observing plan has been developed to prioritize the total data volume returned by THEMIS between the different objectives. This plan devotes 62% of the total data to the IR imager and 38% to the visible, averaged over the course of the Primary Mission. In the baseline plan the IR data will be further sub-divided into 9-wavelength daytime mineralogic observations (47% of total data return) and 3-wavelength nighttime and polar temperature mapping (15% of total data). The visible data will be sub-divided into panchromatic images (37% of total data) and 5-band multi-spectral images (2%). In this example we have assumed a lossless data compression factor of 1.7 for the IR imager and a combination of lossless (40%), and lossy with compression factors of $4\times$ (30%) and $6\times$ (30%) for the visible imager. With these allocations THEMIS will fully map Mars in daytime IR and will map the planet 1.5 times in nighttime IR. The visible imaging will cover $\sim 60\%$ of the planet at 18-m resolution in one band (50,000 18×60km frames) and $< 1\%$ in 5-band color. Tradeoffs between monochromatic and multi-spectra imaging, as well as variations in the degree of lossy compression and pixel summing, will be made to maximize the science return from the visible imager.

5.4. Flight operations

The THEMIS instrument is operated from ASU, building on the facility and staff developed and in place for the MGS TES investigation. No special spacecraft operation or orientation is required to obtain nominal THEMIS data. The instrument alternates between data collection (≤ 3.5 hours per day) and idle modes based on available Deep Space Network (DSN) downlink rates. These modes will fall within the allocated resources (e.g. power), and do not require power cycling of the instrument. All instrument commands are generated at ASU, delivered electronically to the Mars Odyssey Project, and transmitted to the spacecraft. Data are retrieved from the mission database and stored at ASU, where the health and welfare of the instrument are monitored, the data are processed and calibrated, and the archive database is created and stored.

6. Data Reduction and Validation

6.1. OVERVIEW

The data received on the ground are in the form of compressed, scaled, 8-bit 'data numbers' (DN) that represent the delta signal between the scene and the internal reference calibration flag. Data processing consists of decompression, radiometric calibration, and systematic noise removal. For the final generation of geometrically corrected, map-projected data products, each image is processed independently and multiple images can be concatenated together into mosaics. Distortions caused by optics and spacecraft motion are removed concurrently with map projection to avoid multiple resampling degradation.

6.2. DATA CALIBRATION

The THEMIS instrument was radiometrically, spectrally, and spatially calibrated prior to delivery. Three categories of calibration were performed: (1) spatial calibration; (2) spectral calibration; and (3) radiometric calibration including the absolute radiometry, the relative precision (SNR), and the calibration stability during an image collection. The data returned from the instrument in-flight are converted to scene radiance by: (1) adjusting for the gain and offset that were applied in the instrument to optimize the dynamic range and signal resolution for each scene; (2) correcting for drift or offset that occur between observations of the calibration flag; and (3) converting signal to radiance using the instrument response functions determined prior to instrument delivery using a thermal vacuum chamber at the SBRS facility.

Calibration data for the IR sensor were collected at five instrument temperatures (-30, -15, 0, 15, and 30 °C) and seven target temperatures (170, 190, 210, 247, 262, 292, and 307 K). These temperatures were chosen to cover the range of operating and scene temperatures expected at Mars. A full-aperture, calibrated blackbody was placed inside the vacuum chamber and viewed directly by the THEMIS instrument. This blackbody was developed for the SBRS Moderate resolution Imaging Spectrometer (MODIS) instrument, and is a calibrated source traceable to NIST standards with an uncertainty of ± 0.032 K (one sigma). Included in this assessment is a temperature uniformity of 0.020 K, a stability of 0.010 K and an emittance 0.99995 ± 0.00005 (Young, 1999b, c).

For each calibration data set the instrument and blackbody were temperature stabilized at < 0.1 °C. Observations were acquired of the THEMIS internal calibration flag immediately before the collection of a set of calibration images. A 10-sec image was acquired for each point in a matrix of five different gains with five offsets (25 images total) that cover the range of values to be used at Mars. At the completion of this series of images, the calibration flag was reimaged to determine any temperature drift or offset that occurred. The calibration data were adjusted for these minor changes.

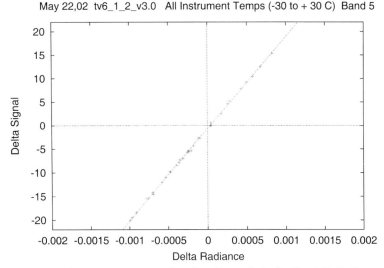

Figure 16. IR imager instrument response function. Example is for Band 5. Radiance units are W cm^{-2} μm^{-1} sr^{-1}.

The response function was computed for each band using the measured output signal versus the computed blackbody radiances. The instrument firmware automatically sets the signal level when viewing the internal calibration flag to 128 DN for a gain of 1, offset of 0. Scene signals are measured relative this value, with the appropriate gain and offset terms applied. As a result, targets colder than the flag have signal DNs (gain and offset removed) <128, whereas warmer scenes have a DN>128.

Figure 16 shows an example of the resulting delta signal (scene-flag; gain = 1, offset = 0) versus delta radiance for Band 5 for all 35 calibration set points. This figure also shows the best-fit linear function that will be used to convert DN to radiance for the Mars data. The measured response values are linear and show no instrument temperature effects. The best-fit response function does not pass through the origin because of the fact that the view of the calibration flag does not include the slightly absorbing fore-optics. Thus, the delta-signal appears slightly negative (less apparent energy) when viewing a scene at the same temperature as the flag. Because the function shown in Figure 16 will be used in flight, this effect is exactly reversible and will not result in any systematic errors in the Mars observations.

The radiometric calibration accuracy of the IR imager is expected to be within $\pm 5\%$; the relative spectral calibration will be to within $\pm 3\%$, which is well within the calibration requirements for mineral spectral analysis.

The THEMIS visible sensor was calibrated over temperature under vacuum by viewing an external calibrated integrating sphere. This sphere was developed for the SBRS Landsat program and is calibrated to a NIST standard with a 1 sigma accuracy of 3.5%. Short term stability is maintained at less than 1% over 14 days (Young, 1999a). Overall uniformity of illumination is better than 0.25% (Young, 1999a).

The visible response function was calculated using a linear fit to the measured signal, with dark current removed, versus calibrated lamp radiance. The measured visible filter functions were integrated over the measured lamp output spectra to produce band-integrated signal versus radiance functions.

6.3. STANDARD DATA PROCESSING

Ground data processing consists of a series of steps that include: depacketization, decompression, instrument health monitoring, radiance calibration, long-term performance monitoring, standard product generation, data quality verification, data archiving and distribution, and preliminary science analysis. These steps must be performed at a rate designed to keep pace with the collection of data by the instrument. The IR data are calibrated to radiance, derived surface kinetic temperature and spectral emissivity, and atmospheric temperature. Geologic processing proceeds with mapping of spectral components, comparison of spectral emissivity with a spectral library, component identification, and identification of temperature anomalies.

6.4. DATA ARCHIVING

Data will be accumulated and archived routinely during standard mission operations as described in the Mars 2001 Odyssey Project Data Archive Plan. All products generated will conform to the standards developed for the Planetary Data System. Data validation and archiving will be performed at ASU. Standard products will consist of images of calibrated radiance in the IR and visible, surface brightness temperature, and atmospheric brightness temperature.

7. Education/outreach

THEMIS investigation builds on the TES Arizona Mars K-12 Education Program, which has over 10 years experience doing highly-leveraged education outreach for Mar Observer and MGS (Edgett, 1995; Edgett and Rice, 1995; Edgett and Christensen, 1996; Edgett et al., 1997). Activities have included: student visits to MGS TES facility; biannual teacher workshops; field trips; workshops at national, regional, and state teacher conferences; visits to schools, community, senior, and science centers; annual teacher guides; lithographs; Spanish material; WWW site; a quarterly newsletter; and a quarterly, age-appropriate, National Standards-based resource for K-8 children.

A major element of the THEMIS outreach program is the THEMIS Student Imaging Project. The objective of this program is to give students at the middle school, high school, undergraduate, and graduate levels the opportunity to participate in the exploration of Mars by planning, acquiring, and analyzing real images of Mars. The THEMIS Student Imaging Program provides background information and teacher mentoring on Mars science to enable classes to plan independent science projects at Mars using THEMIS observations.

Individual classes (grades 5–12) throughout the U.S. research Mars and prepare a short science rational for a target site in conjunction with mentoring and support by the THEMIS staff. The THEMIS staff reviews the student proposals and selects participants. It is expected that approximately 150 classes will participate each year, with high schools selected from each of the 50 states. When possible, selected members of these classes will travel to ASU during the week of image acquisition. The remainder of class will participate in the data acquisition via the Web and through emails, interactive interviews, journals, etc that are designed and implemented by the students. Following the image acquisition, each class will analyze their image in the context of the initial hypothesis- based question that they formulated.

Undergraduate and graduate students or classes participate at a higher level that involves a more detailed specification of a Mars research project. These students prepare a short science rational for a target site and a specific hypothesis to be tested. Students interact with the THEMIS staff to plan image acquisition, and travel to ASU during image acquisition as appropriate or participate via Web interaction. The students analyze their images and participate in a yearly workshop held at ASU. Where appropriate, these projects will be published in the scientific literature.

Acknowledgements

We wish to extend our sincere thanks to all those individuals who have contributed to the detailed design and implementation of this investigation. The highly dedicated team of engineers at Raytheon Santa Barbara Remote Sensing, led by Teresa Fortuna, Martin Greenfield, John Jeffryes, Robert Jensen, Bronislaus Kopra, Chris Laufer, Philip Mayner, Andy Mills, Bill O'Donnell, Paul Owens, Mike Pavlov, George Polchin, and Tom Sprafka, made major contributions to the hardware development. Paul Otjens and Jeff Lewis made major contributions to the VIS engineering at MSSS. Dick Chandos and Spencer Lee provided significant assistance in the final instrument testing and integration. Noel Gorelick, Ben Steinberg, Michael Weiss-Malik, and Saadat Anwar have participated in software development at Arizona State University. Teresa Robinette, Nancy Walizer, and Tara Fisher made major contributions to the administration of this investigation. Carl Kloss provided excellent management of this investigation at the Jet Propulsion Laboratory, and Tim Schofield provided the scientific interface with the Mars Global

Surveyor Project Office. Reviews by Jack Salisbury and two anonymous reviewers significantly improved the manuscript.

References

Bandfield, J. L., 2002, Global mineral distributions on Mars, *J. Geophys. Res.* **107**, 10.1029/2001JE001510.
Bandfield, J. L., Hamilton, V. E. and Christensen, P. R.: 2000a, A global view of Martian volcanic compositions, *Science* **287**, 1626–1630.
Bandfield, J. L., Smith, M. D. and Christensen, P.R.: 2000b, Spectral dataset factor analysis and endmember recovery: Application to analysis of martian atmospheric particulates, *J. Geophys. Res.* **105**, 9573–9588.
Bargar, K. E.: 1978, Geology and thermal history of Mammoth Hot Springs, Yellowstone National Park, Wyoming, *U.S. Geol. Surv. Bull.* **1444**, 55 pp.
Bartholomew, M. J., Kahle, A. B. and Hoover, G.: 1989, Infrared spectroscopy (2.3–20 μm) for the geological interpretation of remotely-sensed multispectral thermal infrared data, *Int. J. Remote Sensing* **10**, 529–544.
Bell, J. F., III, McCord, T. B. and Owensby, P. D.: 1990, Observational evidence of crystalline iron oxides on Mars, *J. Geophys. Res.* **95**, 14447–14461.
Bell, J. F., III, McSween, H. Y., Murchie, S. L., Johnson, J. R., Reid, R., Morris, R. V., Anderson, R. C., Bishop, J. .L., Bridges, N. T., Britt, D. T., Crisp, J. A., Economou, T., Ghosh, A., Greenwood, J. P., Gunnlaugsson, H. P., Hargraves, R. M., Hviid, S., Knudsen, J. M., Madsen, M. B., Moore, H. J., Reider, R., and Soderblom, L.: 2000, Mineralogic and compositional properties of Martian soil and dust: Results from Mars Pathfinder, *J. Geophys. Res.* **105**, 1721–1755.
Blaney, D. L. and McCord, T. B.: 1995, Indications of sulfate minerals in the martian soil from Earth-based spectroscopy, *J. Geophys. Res.* **100**, 14,433–14,441.
Boston, P. J., Ivanov, M. V. and McKay, C. P.: 1992, On the possibility of chemosynthetic ecosystems in subsurface habitats on Mars, *Icarus* **95**, 300–308.
Brock, T. D.: 1978, *Thermophilic Microorganisms and Life at High Temperatures*, Springer-Verlag, New York.
Burns, R. G.: 1993, Origin of electronic spectra of minerals in the visible to near-infrared region, in *Remote Geochemical Analysis: Elemental and Mineralogical Composition*, edited by C.M. Pieters, and P.A.J. Englert, Cambridge University Press.
Carr, M. H.: 1996, *Water on Mars*, Oxford Univ. Press, New York.
Christensen, P. R., Bandfield, J. L., Hamilton, V. E., Howard, D. A., Lane, M. D., Piatek, J. .L., Ruff, S. W., and Stefanov, W. L.: 2000a, 'A thermal emission spectral library of rock forming minerals', *J. Geophys. Res.* **105**, 9735–9738.
Christensen, P. R., Bandfield, J. L., Hamilton, V. E., Ruff, S. W., Kieffer, H. H, Titus, T., Malin, M. C., Morris, R. V., Lane, M. D., Clark, R. N., Jakosky, B. M., Mellon, M. T., Pearl, J. C., Conrath, B. J., Smith, M. D., Clancy, R. T., Kuzmin, R. O., Roush, T., Mehall, G. L., Gorelick, N., Bender, K., Murray, K., Dason, S., Greene, E., Silverman, S. H., and Greenfield, M.: 2001a, 'The Mars Global Surveyor Thermal Emission Spectrometer experiment: Investigation description and surface science results', *J. Geophys. Res.* **106**, 23,823–23,871.
Christensen, P. R., Bandfield, J. L., Smith, M. D., Hamilton, V. E., and Clark, R. N.: 2000b, 'Identification of a basaltic component on the Martian surface from Thermal Emission Spectrometer data', *J. Geophys. Res.* **105**, 9609–9622.
Christensen, P. R., Clark, R. N., Kieffer, H. H., Malin, M. C., Pearl, J. C., Bandfield, J. L., Edgett, K. S., Hamilton, V. E., Hoefen, T., Lane, M. D., Morris, R. V., Pearson, R., Roush, T., Ruff, S. W., and Smith, M. D.: 2000c, 'Detection of crystalline hematite mineralization on Mars by the

Thermal Emission Spectrometer: Evidence for near-surface water', *J. Geophys. Res.* **105**, 9623–9642.

Christensen, P. R. and Harrison, S. T.: 1993, 'Thermal infrared emission spectroscopy of natural surfaces: Application to desert varnish coatings on rocks', *J. Geophys. Res.* **98**(B11), 19,819–19,834.

Christensen, P. R., Malin, M. C., Morris, R. V., Bandfield, J., Lane, M. D., and Edgett, K.: 2000d, 'The distribution of crystalline hematite on Mars from the Thermal Emission Spectrometer: Evidence for liquid water', *Lunar and Planet. Sci.* **XXX**, Abstract # 1627.

Christensen, P. R., Malin, M. C., Morris, R. V., Bandfield, J. L., and Lane, M. D.: 2001b, 'Martian hematite mineral deposits: Remnants of water-driven processes on early Mars, *J. Geophys. Res.* **106**, 23,873–23,885.

Christensen, P. R. and Zurek, R. W.: '1984, Martian north polar hazes and surface ice: Results from the Viking survey/completion mission', *J. Geophys. Res.* **89**, 4587–4596.

Clark, R. N., Swayze, G. A., Singer, R. B., and Pollack, J. B.: 1990, 'High-resolution reflectance spectra of Mars in the 2.3 μm region: Evidence for the mineral scapolite', *J. Geophys. Res.* **95**, 14,463–14,480.

Conel, J. E.: 1969, 'Infrared emissivities of silicates: Experimental results and a cloudy atmosphere model of spectral emission from condensed particulate mediums', *J. Geophys. Res.* **74**, 1614–1634.

Conrath, B., Curran, R., Hanel, R., Kunde, V., Maguire, W., Pearl, J., Pirraglia, J., and Walker, J.: 1973, 'Atmospheric and surface properties of Mars obtained by infrared spectroscopy on Mariner 9', *J. Geophys. Res.* **78**, 4267–4278.

Conrath, B. J., Pearl, J. C., Smith, M. D., Maguire, W. C., Christensen, P. R., Dason, S., and Kaelberer, M.S.: 2000, 'Mars Global Surveyor Thermal Emission Spectrometer (TES) Observations: Atmospheric temperatures during aerobraking and science phasing', *J. Geophys. Res.* **105**, 9509–9520.

Craddock, R. A. and Maxwell, T. A.: 1993, 'Geomorphic evolution of the Martian highlands through ancient fluvial processes', *J. Geophys. Res.* **98**, 3453–3468.

Craddock, R. A., Maxwell, T. A., and Howard, A. D.: 1997, 'Crater morphometry and modification in the Sinus Sabaeus and Margaritifer Sinus regions of Mars', *J. Geophys. Res.* **102**, 13,321–13,340.

Crisp, J., Kahle, A. B., and Abbott, E. A.: 1990, 'Thermal infrared spectral character of Hawaiian basaltic glasses', *J. Geophys. Res.* **95**, 21657–21669.

Davies, D. W., Farmer, C. B., and LaPorte, D. D.: 1977, 'Behavior of volatiles in Mars' polar areas: A model incorporating new experimental data', *J. Geophys. Res.* **82**, 3815–3822.

Edgett, K. S.: 1995, 'To Mars by way of the schoolhouse', *Mercury* **24**, 28–31.

Edgett, K. S. and Christensen, P. R.: 1995, 'Multispectral thermal infrared observations of sediments in volcaniclastic aeolian dune fields: Implications for the Mars Global Surveyor Thermal Emission Spectrometer', *Lunar Planet. Sci.* **XXVI**, 355–356.

Edgett, K. S., Christensen, P. R., Dieck, P. A., Kingsbury, A. R., Kuhlman, S. D., Roberts, J. L., Wakefield, D. A., Rice, J. W. J., and Dodds, J.: 1997, 'K-12 and public outreach for NASA flight projects: Five years (1992–1997) of the Arizona Mars K-12 Education Program', *Lunar Planet. Sci.* **28**, 323–324.

Edgett, K. S. and Rice, J. W. J.: 1995, 'Summary of education and public outreach in Mars Pathfinder Landing Site Workshop II', *LPI Tech. Rept.* **95–01** (Part 2), 17–29.

Edgett, S. K. and Christensen, P. R.: 1996, 'K-12 education outreach program initiated by a university research team for the Mars Global Surveyor Thermal Emission Spectrometer project', *J. Geoscience Education* **44**, 183–188.

Ellis, A. J. and McMahon, W. A. J.: 1977, *Chemistry and Geothermal Systems*, Academic Press, New York.

Exobiology_Working_Group: 1995, An Exobiological Strategy for Mars Exploration, NASA Headquarters.

Farmer, V. C.: 1974, *The Infrared Spectra of Minerals*, 539 pp., Mineralogical Society, London.

Feely, K. C. and Christensen, P. R.: 1999, 'Quantitative compositional analysis using thermal emission spectroscopy: Application to igneous and metamorphic rocks', *J. Geophys. Res.* **104**, 24,195–24,210.

Gaffey, S. J.: 1984, Spectral reflectance of carbonate minerals and rocks in the visible and infrared (0.35 to 2.55 μm) and its applications in carbonate petrology, Ph.D. thesis, University of Hawaii.

Gillespie, A. R., Kahle, A. B., and Palluconi, F. D.: 1984, 'Mapping alluvial fans in Death Valley, CA, using multichannel thermal infrared images', *Geophys. Res. Ltr.* **11**(11), 1153–1156.

Golombek, M. P., Cook, R. A., Moore, H. J., and Parker, T. J.: 1997, 'Selection of the Mars Pathfinder landing site', *J. Geophys. Res.* **102**, 3967–3988.

Greeley, R., Lancaster, N., Lee, S., and Thomas, P.: 1992, Martian Eolian Processes, Sediments, and Features, in *Mars*, edited by H. Kieffer, B. Jakosky, C. Snyder, and M. Matthews, Univ. of Arizona Press, Tucson.

Haberle, R. M. and Jakosky, B. M.: 1990, 'Sublimation and transport of water from the north residual polar cap on Mars', *J. Geophys. Res.* **95**(B2), 1423–1437.

Hamilton, V. E.: 1999, 'Linear deconvolution of mafic igneous rock spectra and implications for interpretation of TES data', *Lunar and Planet. Sci.* **XXX**, CD-ROM, Abstract 1825.

Hamilton, V. E.: 2000, 'Thermal infrared emission spectroscopy of the pyroxene mineral series', *J. Geophys. Res.* **105**, 9701–9716.

Hamilton, V. E. and Christensen, P. R.: 2000, 'Determination of modal mineralogy of mafic and ultramafic igneous rocks using thermal emission spectroscopy', *J. Geophys. Res.* **105**, 9717–9734.

Hamilton, V. E., Wyatt, M. B., McSween, H. Y., and Christensen, P. R.: 2001, 'Analysis of terrestrial and martian volcanic compositions using thermal emission spectroscopy: II. Application to martian surface spectra from MGS TES', *J. Geophys. Res.* **106**, 14,733–14,747.

Hanel, R. A., Conrath, B. J., Hovis, W. A., Kunde, V. G., Lowman, P. D., Pearl, J. C., Prabhakara, C., Schlachman, B., and Levin, G. V.: 1972, 'Infrared spectroscopy experiment on the Mariner 9 mission: Preliminary results', *Science* **175**, 305–308.

Hapke, B.: 1981, 'Bidirectional reflectance spectroscopy I. Theory', *J. Geophys. Res.* **86**, 3039–3054.

Hapke, B.: 1993, Combined Theory of Reflectance and Emittance Spectroscopy, in *Remote Geochemical Analysis: Elemental and Mineralogical Composition*, edited by C. M. Pieters and P. A. J. Englert, Cambridge University Press, Cambridge.

Henderson, B. G., Jakosky, B., and Randall, C. E.: 1992, 'A Monte Carlo Model of Polarized Thermal Emission from Particulate Planetary Surfaces', *Icarus* **99**, 51–62.

Herkenhoff, K. E., and Murray, B. C.: 1990a, 'Color and albedo of the south polar layered deposits on Mars', *J. Geophys. Res.* **95**, 14,511–14,529.

Herkenhoff, K. E. and Murray, B. C.: 1990b, 'High-resolution topography and albedo of the south polar layered deposits on Mars', *J. Geophys. Res.* **95**, 14,511–14,529.

Hook, S. J., Karlstrom, K. E., Miller, C. F., and McCaffrey, K. J. W.: 1994, 'Mapping the Piute Mountains, California, with thermal infrared multispectral scanner (TIMS) images', *J. Geophys. Res.* **99**, 15,605–15,622.

Hunt, G. R. and Logan, L. M.: 1972, 'Variation of single particle mid-infrared emission spectrum with particle size', *Appl. Opt.* **11**, 142–147.

Hunt, G. R. and Salisbury, J. W.: 1970, 'Visible and near-infrared spectra of minerals and rocks: I. Silicate minerals', *Mod. Geol.* **1**, 283–300.

Hunt, G. R., and Salisbury, J. W.: 1976, 'Mid-infrared spectral behavior of metamorphic rocks', *Environ. Res. Paper*, 543-AFCRL-TR-76-0003, 67.

Hunt, G. R. and Vincent, R. K.: 1968, 'The behavior of spectral features in the infrared emission from particulate surfaces of various grain sizes', *J. Geophys. Res.* **73**, 6039–6046.
Jakosky, B. M.: 1998, *The Search for Life on Other Planets*, 336 pp., Cambridge Univ. Press.
Jakosky, B. M., Mellon, M. T., Kieffer, H. H., Christensen, P. R., Varnes, E. S., and Lee, S. W.: 2000, 'The thermal inertia of Mars from the Mars Global Surveyor Thermal Emission Spectrometer', *J. Geophys. Res.* **105**, 9643–9652.
Jakosky, B. M. and Shock, E. L.: 1998, 'The biological potential of Mars, the early Earth, and Europa', *J. Geophys. Res.* **103**, 19359–19364.
James, P. B. and North, G. R.: 1982, 'The seasonal CO2 cycle on Mars: An application of an energy balance climate model', *J. Geophys. Res.* **87**, 10,271–10,284.
Johnson, J. R., Christensen, P. R., and Lucey, P. G.: in press, 'Dust coatings on basalt and implications for thermal infrared spectroscopy of Mars', *J. Geophys. Res.*
Kahle, A., Palluconi, F. D., and Christensen, P. R.: 1993, Thermal emission spectroscopy: Application to Earth and Mars, in *Remote Geochemical Analysis: Elemental and Mineralogical Composition*, edited by C.M. Pieters, and P.A.J. Englert, pp. 99–120, Cambridge University Press, Cambridge.
Kahle, A. B., Madura, D. P., and Soha, J. M.: 1980, 'Middle infrared multispectral aircraft scanner data: Analysis for geological applications', *Appl. Optics* **19**, 2279–2290.
Kieffer, H. H.: 1979, 'Mars south polar spring and summer temperatures: A residual CO2 frost', *J. Geophys. Res.* **84**, 8263–8289.
Kieffer, H. H.: 1990, 'H2O grain size and the amount of dust in Mars' residual north polar cap', *J. Geophys. Res.* **95**, 1481–1494.
Kieffer, H. H., Martin, T. Z., Peterfreund, A. R., Jakosky, B. M., Miner, E. D., and Palluconi, F. D.: 1977, 'Thermal and albedo mapping of Mars during the Viking primary mission', *J. Geophys. Res.* **82**, 4249–4292.
Kieffer, H. H., Titus, T., Mullins, K., and Christensen, P. R.: 2000, 'Mars south polar cap as observed by the Mars Global Surveyor Thermal Emission Spectrometer', *J. Geophys. Res.* **105**, 9653–9700.
Kieffer, H. H. and Zent, A. P.: 1992, 'Quasi-periodic climatic change on Mars, in *Mars*, edited by H. H. Kieffer, B. M. Jakosky, C. W. Snyder, and M. S. Matthews, Univ. of Arizona Press, Tucson.
Komatsu, G. and Baker, V. R.: 1997, 'Paleohydrology and flood geomorphology of Ares Vallis', *J. Geophys. Res.* **102**, 4151–4160.
Lane, M. D. and Christensen, P. R.: 1997, 'Thermal infrared emission spectroscopy of anhydrous carbonates', *J. Geophys. Res.* **102**, 25,581–25,592.
Lazerev, A. N.: 1972, *Vibrational spectra and structure of silicates*, 302 pp., Consultants Bureau, New York.
Lyon, R. J. P.: 1962, Evaluation of infrared spectroscopy for compositional analysis of lunar and planetary soils, in *Stanford Research Institute Final Report Contract NASr*, Stanford Research Institute.
Malin, M. C. and Carr, M. H.: 1999, 'Groundwater formation of martian valleys', *Nature* **397**, 589–591.
Malin, M. C. and Edgett, K. S.: 2000, 'Sedimentary rocks of early Mars', *Science* **290**, 1927–1937.
Malin, M. C. and Edgett, K. S.: 2001, 'Mars Global Surveyor Mars Orbiter Camera: Interplanetary cruise through primary mission', *J. Geophys. Res.* **106**, 23,429–23,570.
Martin, T. Z.: 1986, 'Thermal infrared opacity of the mars atmosphere', *Icarus* **66**, 2–21.
McCord, T. B., Clark, R. N., and Singer, R. B.: 1982, 'Mars: Near-infrared reflectance spectra of surface regions and compositional implications', *J. Geophys. Res.* **87**, 3021–3032.
McKay, D. S., E. K. G. Jr., Thomas-Keprta, K. L., Vali, H., Romanek, C. S., Clemett, S. J., Chillier, X. D. F., Maechling, C. R., and Zare, R. N.: 1996, 'Search for past life on Mars: Possible relic biogenic activity in martian meteorite ALH84001', *Science* **273**, 924–930.

McSween, H. Y., Jr.: 1994, 'What have we learned about Mars from SNC meteorites', *Meteoritics* **29**, 757–779.

Mellon, M. T., Jakosky, B. M., Kieffer, H. H., and Christensen, P. R.: 2000, 'High resolution thermal inertia mapping from the Mars Global Surveyor Thermal Emission Spectrometer', *Icarus* **148**, 437–455.

Moersch, J. E. and Christensen, P. R.: 1995, 'Thermal emission from particulate surfaces: A comparison of scattering models with measured spectra', *J. Geophys. Res.* **100**, 7,465–7,477.

Morris, R. V., Gooding, J. L., Lauer, J. H. V., and Singer, R. B.: 1990, 'Origins of Marslike spectral and magnetic properties of a Hawaiian palagonitic soil', *J. Geophys. Res.* **95**, 14,427–14,435.

Mustard, J. F., Erard, S., Bibring, J.-P., Head, J. W., Hurtrez, S., Langevin, Y., Pieters, C. M., and Sotin, C. J.: 1993, 'The surface of Syrtis Major: Composition of the volcanic substrate and mixing with altered dust and soil', *J. Geophys. Res.* **98**, 3387–3400.

Mustard, J. F. and Hays, J. E.: 1997, 'Effects of hyperfine particles on reflectance spectra from 0.3 to 25 μm', *Icarus* (125), 145–163.

Mustard, J. F. and Sunshine, J. M.: 1995, 'Seeing through the dust: Martian crustal heterogeneity and links to the SNC meteorites', *Science* **267**, 1623–1626.

Nash, D. B. and Salisbury, J. W.: 1991, 'Infrared reflectance spectra of plagioclase feldspars', *Geophys. Res. Lett.* **18**, 1151–1154.

Paige, D. A. and Ingersoll, A. P.: 1985, 'Annual heat balance of martian polar caps: Viking observations', *Science* **228**, 1160–1168.

Palluconi, F. D. and Meeks, G. R.: 1985, Thermal infrared multispectral scanner (TIMS): An investigator's guide to TIMS data, Jet Propulsion Laboratory.

Pearl, J. C., Smith, M. D., Conrath, B. J., Bandfield, J. L., and Christensen, P. R.: 2001, 'Observations of water-ice clouds by the Mars Global Surveyor Thermal Emission Spectrometer experiment: The first martian year', *J. Geophys. Res.* 12,325–12,338.

Pentecost, A.: 1996, High Temperature Ecosystems and their Chemical Interactions with their Environment, in *Evolution of Hydrothermal Ecosystems on Earth (and Mars?)*, edited by G. R. Bock, and J. A. Goode, pp. 99–111, John Wiley and Sons, Chichester.

Pimentel, G. C., Forney, P. B., and Herr, K. C.: 1974, 'Evidence about hydrate and solid water in the martian surface from the 1969 Mariner infrared spectrometer', *J. Geophys. Res.* **79** (No. 11), 1623–1634.

Ramsey, M. S.: 1996, Quantitative Analysis of Geologic Surfaces: A Deconvolution Algorithm for Midinfrared Remote Sensing Data, Ph.D Dissertation thesis, Arizona State University.

Ramsey, M. S. and Christensen, P. R.: 1992, The linear 'un-mixing' of laboratory thermal infrared spectra: Implications for the Thermal Emission Spectrometer (TES) experiment, Mars Observer, *Lunar & Planet. Sci.* **XXIII**, 1127–1128.

Ramsey, M. S. and Christensen, P. R.: 1998, 'Mineral abundance determination: Quantitative deconvolution of thermal emission spectra', *J. Geophys. Res.* **103**, 577–596.

Ramsey, M. S., Christensen, P. R., Lancaster, N., and Howard, D .A.: 1999, 'Identification of sand sources and transport pathways at Kelso Dunes, California using thermal infrared remote sensing', *Geol. Soc. Am. Bull.* **111**, 636–662.

Roush, T. L., Blaney, D. L., and Singer, R. B.: 1993, The surface composition of Mars as inferred from spectroscopic observations, in *Remote Geochemical Analysis: Elemental and Mineralogical Composition*, edited by C. M. Pieters, and P. A. J. Englert, Cambridge University Press.

Ruff, S. W. and Christensen, P. R.: 2002, 'Bright and dark regions on Mars: Particle size and mineralogical characteristics based on Thermal Emission Spectrometer data', *J. Geophys. Res.*, in press.

Salisbury, J. W.: 1993, Mid-infrared spectroscopy: Laboratory data, in *Remote Geochemical Analysis: Elemental and Mineralogical Composition*, edited by C. Pieters, and P. Englert, pp. Ch. 4, Cambridge University Press, Cambridge.

Salisbury, J. W., D'Aria, D. M., and Jarosewich, E.: 1991, 'Mid-infrared (2.5–13.5 um) reflectance spectra of powdered stony meteorites', *Icarus* **92**, 280–297.

Salisbury, J. W. and Eastes, J. W.: 1985, 'The effect of particle size and porosity on spectral contrast in the mid-infrared', *Icarus* **64**, 586–588.

Salisbury, J. W., Hapke, B., and Eastes, J. W.: 1987a, 'Usefulness of weak bands in midinfrared remote sensing of particulate planetary surfaces', *J. Geophys. Res.* **92**, 702–710.

Salisbury, J. W. and Wald, A.: 1992, 'The role of volume scattering in reducing spectral contrast of reststrahlen bands in spectra of powdered minerals', *Icarus* **96**, 121–128.

Salisbury, J. W., Wald, A., and D'Aria, D. M.: 1994, 'Thermal-infrared remote sensing and Kirchhoff's law 1. Laboratory measurements', *J. Geophys. Res.* **99**, 11897–11911.

Salisbury, J. W. and Walter, L. S.: 1989, 'Thermal infrared (2.5–13.5 μm) spectroscopic remote sensing of igneous rock types on particulate planetary surfaces', *J. Geophys. Res.* **94**(No. B7), 9192–9202.

Salisbury, J. W., Walter, L. S., and Vergo, N.: 1987b, Mid-infrared (2.1–25 μm) spectra of minerals: First Edition, in *U.S.G.S., Open File Report*, pp. 87–263, U. S. Geological Survey Open File Report.

Selivanov, A. S., Naraeva, M. K., Panfilov, A. S., Gektin, Y. M., Kharlamov, V. D., Romanov, A. V., Fomin, D. A., and Miroshnichenko, Y. Y.: 1989, 'Thermal imaging of the surface of Mars', *Nature* **341**, 593–595.

Shock, E. L.: 1997, 'High temperature life without photosynthesis as a model for Mars', *J. Geophys. Res.*

Shoemaker, E. M.: 1963, Impact mechanics at Meteor Crater, Arizona, in *The Moon, Meteorites, and Comets*, edited by B. M. Middlehurst, and G. P. Kuiper, pp. 301–336, Univ. of Chicago Press, Chicago.

Silverman, S., Bates, D., Schueler, C., O'Donnell, B., Christensen, P., Mehall, G., Tourville, T. and Cannon, G.: 1999, Miniature Thermal Emission Spectrometer for the Mars 2001 Lander, *Proceedings of the IEEE*.

Singer, R. B.: 1982, 'Spectral evidence for the mineralogy of high-albedo soils and dust on Mars', *J. Geophys. Res.* **87**, 10,159–10,168.

Smith, M. D., Bandfield, J. L., and Christensen, P. R.: 2000, 'Separation of atmospheric and surface spectral features in Mars Global Surveyor Thermal Emission Spectrometer (TES) spectra: Models and atmospheric properties', *J. Geophys. Res.* **105**, 9589–9608.

Smith, M. D., Conrath, B. J., Pearl, J. C., and Christensen, P. R.: 2002, 'Thermal Emission Spectrometer observations of martian planet-encircling dust storm 2001A', *Icarus* **157**, 259–263.

Smith, M. D., Pearl, J. C., Conrath, B. J., and Christensen, P. R.: 2001a, 'Thermal Emission Spectrometer results: Mars atmospheric thermal structure and aerosol distribution', *J. Geophys. Res.* **106**, 23929–23945.

Smith, M. D., Pearl, J. C., Conrath, B. J., and Christensen, P. R.: 2001b, 'One Martian year of atmospheric observations by the Thermal Emission Spectrometer', *Geophys. Res. Letters* **28**, 4263–4266.

Smith, M. D., Pearl, J. C., Conrath, B. J., and Christensen, P. R.: 2001c, 'Thermal Emission Spectrometer results: Atmospheric thermal structure and aerosol distribution', *J. Geophys. Res.* **106**, 23,929–23,945.

Stevens, T. O. and McKinley, J. P.: 1995, 'Lithoautotrophic microbial ecosystems in deep basalt aquifers', *Science* **270**, 450–454.

Tamppari, L. K., Zurek, R. W., and Paige, D. A.: 2000, 'Viking era water ice clouds', *J. Geophys. Res.* **105**, 4087–4107.

Tanaka, K. L.: 1997, 'Sedimentary history and mass flow structures of Chryse and Acidalia Planitiae', Mars, *J. Geophys. Res.* **102**, 4131–4149.

Tanaka, K. L. and Leonard, G. J.: 1995, 'Geology and landscape evolution of the Hellas region of Mars', *J. Geophys. Res.* **100**, 5407–5432.

Thomas, P. and Gierasch, P. J.: 1995, 'Polar margin dunes and winds on Mars', *J. Geophys. Res.* **100**, 5379–5406.

Thomas, P. C., Malin, M. C., and Edgett, K. S.: 2000, 'North-south geological differences between the residual polar caps on Mars', *Nature* **404**, 161–164.

Thomas, P. C., Squyres, S., Herkenhoff, K., Howard, A., and Murray, B.: 1992, Polar deposits of Mars in Mars, in *Mars*, edited by H. Kieffer, B. Jakosky, C. Snyder, and M. Matthews, pp. 767–795, Univ. Arizona Press, Tucson.

Thomas, P. C. and Weitz, C.: 1989, 'Sand dune materials and polar layered deposits on Mars', *Icarus* **81**, 185–215.

Thomson, J. L. and Salisbury, J. W.: 1993, 'The mid-infrared reflectance of mineral mixtures (7–14 μm)', *Remote Sensing of Environment* **45**, 1–13.

Titus, T. N., Kieffer, H. H., Mullins, K. F., and Christensen, P. R.: 2001, 'TES Pre-mapping data: Slab ice and snow flurries in the Martian north polar night', *J. Geophys. Res.* **106**, 23,181–23,196.

Van der Marel, H. W. and Beeutelspacher, H.: 1976, Atlas of infrared spectroscopy of clay minerals and their admixtures, 396pp., Elsevier Scientific Publishing Co., Amsterdam.

Vincent, R. K. and Hunt, G. R.: 1968, 'Infrared reflectance from mat surfaces', *Appl. Opt.* **7**, 53–59.

Vincent, R. K. and Thompson, F.: 1972, 'Spectral compositional imaging of silicate rocks', *J. Geophys. Res.* **17** (No. 14), 2465–2472.

Wald, A. E. and Salisbury, J. W.: 1995, 'Thermal infrared emissivity of powdered quartz', *J. Geophys. Res.* **100**, 24665–24675.

Walter, M. R. and Des Marais, D. J.: 1993, 'Preservation of biological information in thermal spring deposits: Developing a strategy for the search for fossil life on Mars', *Icarus* **101**, 129–143.

White, D. E., Hutchinson, R. A., and Keith, T. E. C.: 1988, *The Geology and Remarkable Thermal Acitivty of Norris Geyser Basin, Yellowstone National Park*, U.S. Geol. Survey Prof. Paper 1456.

Wilson, E. B., Jr., Decius, J. C., and Cross, P. C.: 1955, *Molecular Vibrations: The Theory of Infrared and Raman Vibrational Spectra*, McGraw-Hill.

Wyatt, M. B.,, Hamilton, V. E., McSween, J. H. Y., Christensen, P. R., and Taylor, L. A.: 2001, 'Analysis of terrestrial and martian volcanic compositions using thermal emission spectroscopy: I. Determination of mineralogy, chemistry, and classification strategies', *J. Geophys. Res.* **106**, 14,711–14,732.

Young, J.: 1999a, FM1 Absolute radiometric calibration reflectance region: uncertainty estimate, Raytheon Santa Barbara Remote Sensing, Santa Barbara.

Young, J.: 1999b, FM1 Absolute radiometric calibration thermal region: uncertainty estimate, Raytheon Santa Barbara Remote Sensing, Santa Barbara.

Young, J.: 1999c, Summary of the Blackbody Calibration Source (BCS) and Space View Source (SVS) refurbishment and calibration process, Raytheon Santa Barbara Remote Sensing, Santa Barbara.

MARTIAN RADIATION ENVIRONMENT EXPERIMENT (MARIE)

GAUTAM D. BADHWAR
Earth and Solar System Exploration Division
NASA Johnson Space Center, Houston, Texas 77058-3696
(Author for correspondence, E-mail: Francis A. Cuccinotta1@jsc.nasa.gov)

(Received 11 January 2001; Accepted in final form 12 February 2003)

Abstract. Measurements of radiation levels at Mars including the contributions of protons, neutrons, and heavy ions, are pre-requisites for human exploration. The MARIE experiment on the Mars-01 Odyssey spacecraft consists of a spectrometer to make such measurements in Mars orbit. MARIE is measuring the galactic cosmic ray energy spectra during the maximum of the 24th solar cycle, and studying the dynamics of solar particle events and their radial dependence in orbit of Mars. The MARIE spectrometer is designed to measure the energy spectrum from 15 to 500 MeV/n, and when combined other space based instruments, such as the Advanced Composition Explorer (ACE), would provide accurate GCR spectra. Similarly, observations of solar energetic particles can be combined with observations at different points in the inner heliosphere from, for example, the Solar Heliospheric Observatory (SOHO), to gain information on the propagation and radial dependence in the Earth-Mars space. Measurements can be compared with the best available radiation environment and transport models in order to improve these models for subsequent use, and to provide key inputs for the engineering of spacecraft to better protect the human crews exploring Mars.

1. Introduction

In deep space there are two sources of radiation of concern for manned spaceflight; galactic cosmic rays (GCR), and solar energetic particles (SEP). The steady low dose rate from GCR particles leads to risks of late effects, such as cancer, cataracts and damage to the central nervous system (NCRP, 2002). Passage of heavy nuclei through the brain may lead to central nervous system damage that may result, for example, in memory loss, and potentially impact a human mission to Mars. The GCR contains fully stripped ions from hydrogen to uranium and span an energy range from ~ 1 MeV/n to 10^{14} MeV/n (Simpson, 1983). However, the flux of ions above nickel is too small to be of concern and energies above ~ 10 GeV/n do not contribute to the radiation dose. The intensity of GCR is inversely related to solar activity, but this source of radiation is always present and because of the high energy of particles, difficult to shield against. Solar energetic particles (SEP) are sporadic events leading to sharp short-term increase in dose and dose rate possibly causing harmful acute radiation effects and adding risks of late effects. The 2001 Mars Odyssey mission will be launched around the peak of the current 24th solar cycle. Analysis of previous SEP data (Feynman *et al.*, 1990) shows that almost all of the SEP events fall between about 3 years prior and 4 years past the solar

TABLE I

Recommended dose equivalent limits (Sv) for low-earth orbit flights (NCRP-132, 2000)

	BFO	Skin	Ocular lens
Career	0.5–3.0	6	4
Annual	0.5	3	2
30 day	0.25	1.5	1

maximum in the ~ 11 year solar cycle. Nymmik (1999) showed that the number of SEP events, N per year, with a fluence, F greater than 10^5 protons cm^{-2}, is given by: $N = 0.3W^{0.75}$, where W is the smoothed Wolf number relate to the total number of sunspots and sunspot groups. A number of events in the current solar cycle have been observed (for e.g., November 1997, July 2000, and November 2000) in earth orbit and contributed to the exposure of the astronauts in the Mir and the International Space Station. However in Earth orbit the Earth's magnetic shielding provides a great reduction in radiation exposures, and magnetic shielding is not present in transit to Mars or on the Martian surface. Such events would, of course, generate considerably higher doses in free space.

Radiation exposure limits for interplanetary missions have not been defined at this time. Low-earth orbit (LEO) limits are given in Table 1 and are used for guidance for Mars mission design studies. The career limits followed in low Earth orbit are based on estimates of the dose that would lead to a 3% increase excess cancer mortality (NCRP, 2000). The annual and 30-day limits in Table 1 are intended to prevent the occurrence of any clinical significant health risks, including acute radiation sickness, and are most important when considering SEP events.

Interest in a possible human Mars mission in the next several decades has remained high. The National Academy of Sciences (1993) reported that radiation protection posed a significant challenge to the design of such missions. Significant progress in the development of models of the galactic cosmic radiation environment (Badhwar and O'Neill, 1996, Nymmik, 1996), improvements were made, which have been implemented into models for predicting the dose rate during the cruise phase (Badhwar, 1994), most notably the NASA Langley Research Center transport models BRYNTRN and HZETRN (Wilson et al., 1995). Simonsen et al., (2000), has used the HZETRN model to carry out detailed analysis of expected dose rates for a potential Mars mission. A mission with a 1-year transit time and 1.5 year surface stay has been under consideration as a possible first human flight. Simonsen et al., (2000) using the recently developed concept of a lightweight inflatable habitable (denoted as TransHab) volume calculated the expected exposures

TABLE II
Dose equivalent estimates (in units cSv = 1 rem) for a potential Mars mission.
(Simonsen et al., 2000).

Source	1 yr Transit Dose Equivalent		1.5 yr Surface Dose Equivalent	
	Skin	BFO	Skin	BFO
Solar Maximum	33.4	27.0	20.1	17.6
Solar Minimum	93.8	72.7	46.5	40.7
August 1972 SEP	63.8	17.0	4.6	2.4

on such a mission. Table 2 gives their results where the dose equivalent to the blood forming organs (BFO) and skin are estimated.

Uncertainties in the GCR model predictions and errors in the fragmentation cross-sections and their energy dependence directly impact the radiation dose and design of an appropriate shielding. The MARIE experiment will provide valuable data on the GCR environment in Mars orbit and in transit from Mars to Earth. Ground-based accelerator studies are considered for reducing errors in fragmentation cross-sections and transport codes. A study by Badhwar et al., (1994) indicated that approximately 17.5 g cm^{-2} of aluminum equivalent material would be sufficient to provide shielding during a solar minimum transit phase to stay below the 50 cSv annual BFO limit. New calculations, however, with an improved version of the HZETRN code show that almost 50 g cm^{-2} is required for the same level of protection. This is due to the flattening of the dose at large aluminum shielding thickness (Simonsen et al., 2000). Thus, the new challenge is to choose low atomic weight shielding materials and to significantly reduce the errors in estimating radiation exposures.

In estimating crew radiation exposures, the contribution of the secondary neutrons is currently not fully understood. This is a more serious problem with aluminum type shielding than with carbon based or more hydrogenated materials. It will definitely be of concern for a Lander mission due to production of neutrons in the Martian atmosphere and from the neutrons back scattered of the Mars soil or atmosphere. Since only an orbiter instrument will be part of the Mars'01 mission, the focus will be to determine the galactic cosmic ray energy spectra during the maximum of the 24th solar cycle, and study the dynamics of SEP events and their radial dependence, during the cruise phase and during the orbit phase around Mars.

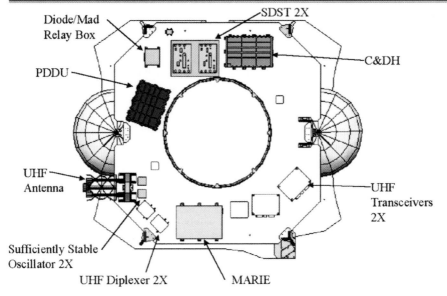

Figure 1. Location of MARIE at launch configuration.

2. MARIE Instrumentation

The MARIE instrument consists of an energetic particle spectrometer that can measure the elemental energy spectra of charged particles of the energy range of 15–500 MeV/n. These limits are charge dependent and would extend to about 600 MeV/n for iron nuclei. The spectrometer is mounted on the science deck (Figure 1) with an angular acceptance of 50°. As the spacecraft orbits Mars, the axis of this field of view sweeps a cone or swath of directions on the sky outwards from Mars. During each orbit, the angle between the axis of the spectrometer's field of view and the mean interplanetary field direction varies from 0° to 180°.

Figure 2 shows a schematic of the spectrometer. It consists of a set of solid-state detectors and a high refractive index Cherenkov detector. The spectrometer geometry is defined by two $25.4 \times 25.4 \times 1$ mm thick ion-implanted silicon solid-state detectors A1 and A2 that are operated at 160 V. In between A1 and A2 there are two 24×24 mm position sensitive detectors PSD1 and PSD2, each with a 24×24 mm wire grid, to define the incident direction of charged particle with respect to the axis of the telescope. The A2 detector is followed by a set of 5 mm thick lithium-drifted silicon solid-state detectors (B1, B2, B3, and B4), and a high refractive index Schot-glass Cherenkov (C) detector. The spectrometer is built like a personal computer. Each detector has its own card, with all of the electronics associated with the detector on it, including a 12-bit analog-to-digital (ADC) converter, and Field

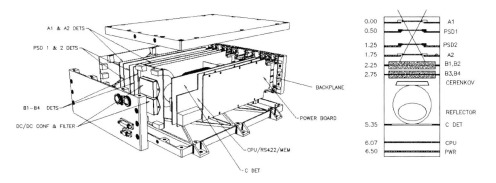

MARIE ORBITER

Figure 2. Schematic of Orbiter Spectrometer (Dimension in Inches) with scale information shown on the left and a schematic of the series of silicon detectors shown on the right.

Programmable Gate Array (FPGA). The main power supply is a nominal 28 V (16–32) DC Interpoint unit. There is an 80 MB flash memory for data storage. The CPU board has an Intel microprocessor, and data communication hardware for transferring data through RS 422 and RS 232 ports (9600 baud). The memory device can store data for more than two weeks of operations. The instrument is a dE/dx x E spectrometer, where E is the energy of the particle and dE/dx the energy loss per unit path length, for stopping particles and dE/dx x C spectrometer for penetrating particles, where C is a constant. Following coincidence rates are also recorded: A1A2, A1A2B1, A1A2B1B2, A1A2B1B2B3, A1A2B1B2B3B4, and A1A2B1B2B3B4C. The basic trigger is the A1A2 coincidence and requires a proton with energy > 15 MeV. Following such a trigger all of the detectors are read out.

Table 3 gives the area-solid angle product for various coincidence rates assuming an isotropic angular distribution of incident particles. An isotropic distribution follows a $\cos^2(\theta)$ distribution and sea level muons are known to follow a $\cos^4(\theta)$ distribution. The comparison in Table 3 indicate that the detector is performing correctly. Since the PSD detectors are slightly smaller than the trigger A detectors, about 15% of the particles miss these detectors and incident angle information for such events is lost. These detectors have separate thresholds for rows and columns.

If two wires have signals above the threshold, the position of each wire is noted; however, if more than two wires have signals above the threshold, only the median position of the wires is recorded. In addition, the magnitude of the total charge from the whole detector is read-out through a 10-bit ADC. Thus the PSDs not only provide the (x, y) coordinates of the passage of the particle through the detector, but a measure of the ionization loss in the detectors also. Based on the 1 mm wire spacing and separation of the two detectors, the incident angle of the particle can be measured to better than ±3%.

TABLE III

Area-solid product for an isotropic and sea-level muon angular distribution

Coincidence	Area-solid angle (cm² sr) –isotropic	Area-solid angle (cm² sr) –sea-level
A1A2	1.0	1.0
A1PSD1PSD2A2	0.85	0.85
A1A2B1	1.0	1.0
A1A2B1B2	0.996	0.997
A1A2B1B2B3	0.958	0.966
A1A2B1B2B3B4	0.907	0.922

Figure 3 is a typical example of the pulse height distribution in various silicon detectors for sea-level muons. They have the characteristic Landau distribution. The width of the distribution arises from the quadratic sum of the detector noise, fluctuations in the ionization energy loss, the Landau-Vavilov distribution, and the variation in path length in the detector.

Using the incident direction derived from the PSD detectors these data can be corrected for the acceptance angle leading to better charge and energy resolution. Figure 4 shows an example for the sea-level muons. The full-width at half-maximum for the Landau-Vavilov distribution decreases approximately inversely as a function of increasing charge and is also related to the overall charge resolution of the instrument.

The dose, D, and dose equivalent, H, rates needed for calculating astronaut exposures are given by:

$$D = (k/\rho) \int LJ(L)dL$$

$$H = (k/\rho) \int LQ(L)J(L)dL$$

Here $J(L)$ is the differential linear energy transfer (LET; $L \equiv dE/dx$) spectrum of the incident particles, Q is the radiation quality factor as a function of the linear energy transfer, ρ the density of tissue, and k a multiplier for appropriate definition of the radiation units. Using the two trigger detectors A1 and A2 the LET distribution can be measured directly assuming that particles traverse the detectors at the mean incident angle. A more accurate distribution can be obtained using the knowledge of incident angles from the PSDs. Thus, direct information for the human exploration missions can be obtained if just the two detectors function normally. An example of the LET distribution measured with a silicon based LET spectrometer in a 51.65°

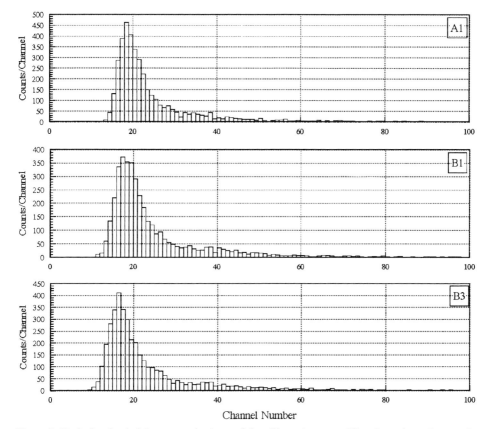

Figure 3. Typical pulse height spectra in three of the silicon detectors. The channel number can be related to energy loss of a particle in the detector.

inclination orbit Space Shuttle flight (STS-91) around the time of solar minimum (1996) and comparison with HZETRN model calculations is shown in Figure 5.

Also, the knowledge of energy loss in all the detectors can be used to determine the actual energy spectra and make a more accurate assessment of individual charges.

Figure 6 plots the sea-level muon spectra in the Cherenkov detector, C, when the detector is accepting particles from the top and when the telescope has been inverted. Most of the photons coming out of the Schot-glass follow the direction of the incident particle. If the telescope is inverted, the photo-multiplier tube can no longer view these photons. Thus the signal in C is very weak. This provides an ability to distinguish the direction of the high-energy particles.

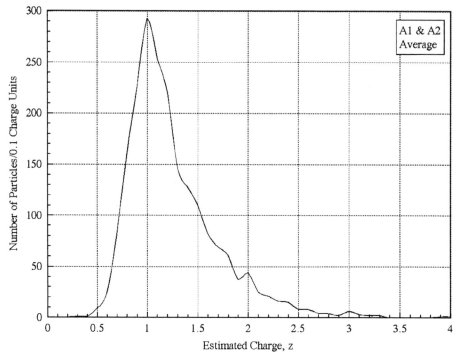

Figure 4. Calculated charge distribution using data for A1 and A2 detectors for muons (Z = 1)

3. Data Analysis

Data from the spectrometer can be analyzed as follows. The data are divided into three energy regimes: (1) particles that stop in any detector from A2 to B4, (2) particles that go through B4 but do not give a Cherenkov signal, and (3) particles that give a Cherenkov signal. The angular information from the PSD detector would ensure that the particle falls in the right geometry such that the particle energy and charge is correctly identified.

Case 1: If the range energy relation for protons in silicon is expressed as a power law in energy with index, n, $R = K E^n$, and ΔE the energy loss in the thin detectors (A1 or A2). If E is the total energy deposited by the stopping particle, then

$$[\Delta E \times (E - \Delta E)] \propto (Z^2 M^{n-1})$$

where Z is the charge number and M is the mass of the incident particle, quantities that are element and isotope specific.

Thus, for each isotope, there exists a separate hyperbolic curve. Charge, mass and energy of the stopping particle can be determined. Figure 7 shows a plot of the energy loss in A1 versus the energy loss in A2 for protons and helium nuclei from the flight of a similar telescope (with no PSDs but an anti-coincidence counter).

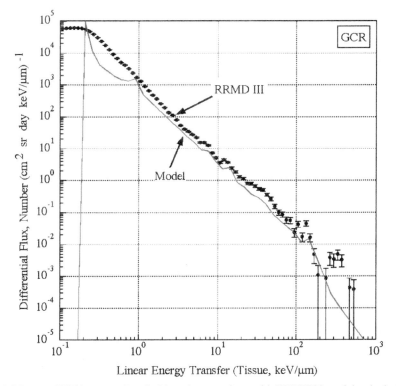

Figure 5. Measured LET spectra (symbols) and comparison with HZETRN model calculations for GCR Component (line).

Case 2: In this case, the particle energy is assumed to be greater than the energy required to penetrate all of the detectors but less than the Cherenkov threshold of ∼ 180 MeV/n. The measurements of energy loss in each of the six (and in most cases eight) silicon detectors are compared with values calculated for a given energy. The energy is varied to find the best fit by minimizing the merit function:

$$\chi^2(E) = \sum [\Delta E_i^{cal} - \Delta E_i^{obs}]^2$$

where i (= 1,6) is the index for the detector and 'cal' and 'obs' refer to the calculated - and observed-energy loss.

Case 3: In this case, the light emitted in the Cherenkov detector, C, is given by:

$$C = kZ^2(1 - \beta_o^2/\beta^2)$$

where β_o is threshold velocity (= 1/real part of index of refraction), and β is the particle velocity and the energy loss, ΔE, is

$$(\Delta E) \propto (Z^2/\beta^2)$$

These two equations are solved to obtain the charge and velocity (energy per nucleon). Using Schot-glass we can cover the energy range from about 180 MeV/n to 500 MeV/n.

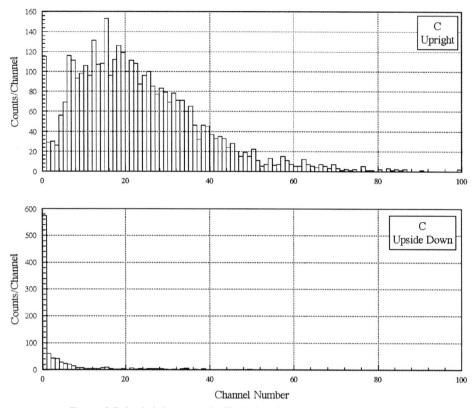

Figure 6. Pulse height spectra in Cherenkov detector in two orientations.

Thus, this technique can cover the entire range from the minimum energy required to form the A1A2 coincidence, to the energy where the Cherenkov response saturates. In addition, one can construct an integral spectrum of SEP events, independently, from various coincident rates. An additional advantage of such an instrument is that using the energy loss in A1 and A2 and the angle of the incident particle, a *true* LET spectrum of particles can be determined.

The area solid angle of the instrument requires large time integration in order to assure quality statistics. We plan to use our measurements of protons and helium GCR particles and correlate them with measurements from the ACE instrument (at L1, Earth's Lagrange point), to obtain accurate GCR spectra in the Martian orbit.

The GCR are isotropic in free-space, and the SEP can have specific angular distributions dependent on the magnetic field-lines that connect the sun to Mars. The data from MARIE for the GCR and SEP events measured can be used as input spectral functions to radiation transport models to predict the flux, dose, and LET spectra on the Martian surface. Because the arrival direction and distribution of SEP's can be measured, proper account can be taken of the shielding presented by the Martian atmosphere and the solid planet to different arrival directions. These

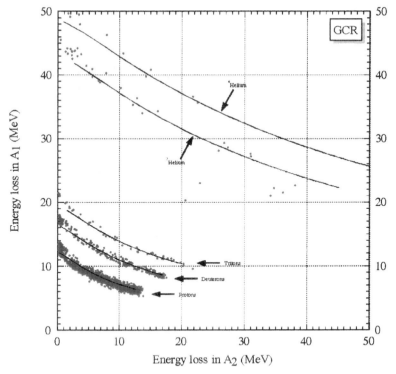

Figure 7. Energy loss curves determined for various particles.

predicted fluxes, doses and LET spectra can be compared with measurements made by a future Lander instrument to determine the radiation transport characteristics of the Martian atmosphere and provide key radiation information for both the cruise and landed phases of a future manned interplanetary mission. Data on the gamma-rays and neutrons scattered of the surface and atmosphere of Mars are being collected by the GRS and HEND experiments on Odyssey (Feldman *et al.*, 2002; Mitrofanov *et al.*, 2002). Such data could be correlated with the primary GCR composition measured by MARIE and radiation transport models, to provide interesting tests of our knowledge of radiation transport on Mars. Excellent tools for such predictions have been recently been reported by Wilson *et al.* (1999) and Reddy and Howe (1999).

4. Conclusions

The Orbiter spectrometer will provide data on GCR and SEP from the cruise phase to Martian orbit and in the Martian orbit. The information can be used for future human missions to Mars providing validation of models to be used by mission planners for estimating dose, dose equivalent and LET-spectra on such missions. The measurements of the GCR charge and energy spectra will provide important information on possible differences in solar modulation of the GCR near-Earth and at Mars. Finally, we anticipate that the observation of one or more SEP events at Mars or on Mars transit will provide valuable data for estimating risks and validating methods for future human exploration missions.

References

Badhwar, G. D., and O'Neill, P. M.: 1996, 'Galactic cosmic radiation model and its applications', *Adv. Space Res.* **17(10)**, 7–17.

Badhwar, G. D., Cucinotta, F. A., and O'Neill, P. M.: 1994, 'An analysis of interplanetary space radiation exposure for various solar cycles,' *Radiat. Res.* **138**, 201–208.

Feldman *et al.*: 2002, 'Global Distribution of Neutrons from Mars: Results from Mars Odyssey', *Science* **297**, 75–78.

Feynman, J., Armstrong, T. P., Dao-Gibner, L., and Silverman, S.: 1990, 'New interplanetary proton fluence model', *J. Spacecraft* **27**, 403–410.

Mitrofanov, I. *et al.*: 2002, 'Maps of Subsurface Hydrogen fron the High Energy Neutron Detector, Mars Odyssey', *Science* **297**, 78–81.

National Council on Radiation Protection and Measurements: 2002, Radiation Protection Guidance for Activities in Low-Earth Orbit. Bethesda, MD: National Council on Radiation Protection and Measurements: NCRP Report No. 132.

Nymmik, R. A.: 1999, 'The problems of cosmic ray particle simulation for the near-Earth orbital and interplanetary flight conditions', *Rad. Meas.* **30**, 669–677.

Reddy, R. C. and Howe, S. D.: 1999, The Martian Radiation Environment. Workshop on Mars 2001: Integrated Science in *Preparation for Sample Return and Human Exploration* (J. Marshall and C. Weitz, eds.), LPI Contribution No. 991 (Lunar and Planetary Institute, Houston), 87–99, Houston Texas.

Simonsen, L. C., Wilson, J. W., Kim, M. H., and Cucinotta, F. A.: 2000, 'Radiation exposure for human Mars exploration', *Health Physics* **79**, 515–525.

Simpson, J. A.: 1983, 'Elemental and isotopic composition of the galactic cosmic rays', *Ann. Rev. Nucl. Part. Sci.* **33**, 323–381.

Space Studies Board, National Research Council: 1993, Scientific Prerequisites for Human Exploration, National Academy Press, Washington, D.C..

Wilson, J. W., Badavi, F. F., Cucinotta, F. A., Shinn, J. L., Badhwar, G. D., Silberberg, R., Tsao, C. H., Townsend, L. W., and Tripathi, R. K.: 1995, HZETRN: description of a free-space ion and nucleon transport and shielding computer program. Springfield, VA: National Technical Information Service: NASA TP 3495 .

Wilson, J. W., Kim, M. Y., Clowdsley, M. S., Heinbockel, J. H., Tripathi, R. K., Singleterry, R. C., Shinn, J. L., and Suggs, R.: 1999, Mars Surface Ionizing Radiation Environment, Workshop on Mars 2001: Integrated Science in Preparation for Sample Return and Human Exploration (J. Marshall and C. Weitz, eds.), LPI Contribution No. 991 (Lunar and Planetary Institute, Houston), 112–114, Houston Texas.

RADIATION CLIMATE MAP FOR ANALYZING RISKS TO ASTRONAUTS ON THE MARS SURFACE FROM GALACTIC COSMIC RAYS

PREMKUMAR B. SAGANTI[1], FRANCIS A. CUCINOTTA[2], JOHN W. WILSON[3], LISA C. SIMONSEN[3] and CARY ZEITLIN[4]

[1]Lockheed Martin Space Operations, Houston TX-77058, U.S.A.;
[2]NASA Johnson Space Center, Houston, TX-77058, U.S.A.;
[3]NASA Langley Research Center, Hampton, VA-23681, U.S.A.
[4]Lawrence Berkeley National Laboratory, University of California, Berkeley, CA-94720, U.S.A.
(*Author for correspondence, currently with the NASA Center for Applied Radiation Research, A+M University, Prairie View, TX-77446, USA, E-mail: premkumeur@saqanti1@jsc.nasa.gov)

(Received 14 January 2002; Accepted in final form 16 January 2003)

Abstract. The potential risks for late effects including cancer, cataracts, and neurological disorders due to exposures to the galactic cosmic rays (GCR) is a large concern for the human exploration of Mars. Physical models are needed to project the radiation exposures to be received by astronauts in transit to Mars and on the Mars surface, including the understanding of the modification of the GCR by the Martian atmosphere and identifying shielding optimization approaches. The Mars Global Surveyor (MGS) mission has been collecting Martian surface topographical data with the Mars Orbiter Laser Altimeter (MOLA). Here we present calculations of radiation climate maps of the surface of Mars using the MOLA data, the radiation transport model HZETRN (high charge and high energy transport), and the quantum multiple scattering fragmentation model, QMSFRG. Organ doses and the average number of particle hits per cell nucleus from GCR components (protons, heavy ions, and neutrons) are evaluated as a function of the altitude on the Martian surface. Approaches to improve the accuracy of the radiation climate map, presented here using data from the 2001 *Mars Odyssey* mission, are discussed.

1. Introduction

The potential for harmful late effects including cancer, cataracts, neurological disorders, and non-cancer mortality risks, from galactic cosmic rays (GCR) pose a major threat for the human exploration of Mars (National Academy of Sciences Space Science Board, 1973; National Academy of Sciences, 1997; Cucinotta *et al.*, 2001). Because of their high energies, the GCR are extremely penetrating and cannot be eliminated by practical amounts of shielding (Wilson *et al.*, 1991; Cucinotta, *et al.*, 2001). The high charge and energy HZE ions portion of the GCR present unique challenges to biological systems such as DNA, cells, and tissue and the risks to humans is highly uncertain at this time (National Academy of Sciences, 1997; Cucinotta *et al.*, 2001). Another threat in deep space is solar particle events, which could induce acute radiation syndromes including death if the event is large enough and increase the risk of cancer or other late effects. However, shielding offered by the Mars atmosphere and spacecraft structures along with the early warning and detection systems may be effective as mitigation measures. The biological effects of protons are well understood being similar to gamma rays. Safety assurances and dose limits for the human exploration of Mars cannot be provided at this time due to the uncertainties in the biological effects of HZE ions (National Academy of Sciences, 1997). Important physical data on the GCR elemental and energy composition near Mars and on the Mars surface will be needed prior to human exploration. Developing methods for the accurate prediction of the modulation of the isotopic composition and energies of the GCR after nuclear and atomic interactions with Mars atmosphere and soil and the accurate determination of secondary neutron spectra will be essential to design and undertake such missions.

Robotic precursor missions to Mars can provide valuable data on the radiation environment to be encountered in future human exploration missions of the red planet and validation of models used for mission design. These measurements should include direct physical measurements of the energy spectra of protons, heavy ions, and neutrons with radiation spectrometers. Physical data related to the Mars altitude (Smith et al., 1999), atmospheric, and soil composition will also be valuable in developing models of astronaut and equipment exposures, and in designing shielded habitat configurations. In this paper, we discuss the use of the Mars Orbiter Laser Altimeter (MOLA) altitude data (Smith et al., 1999), and models of the radiation environment and GCR transport to present the first complete radiation map of the Mars surface.

2. Methods

The HZETRN (High Z and Energy Transport) code (Wilson et al., 1991; Clowdsley et al., 2001) describes the atomic and nuclear reaction processes that alter the GCR in their passage through materials such as the Mars atmosphere and tissue. The HZETRN code solves the Boltzmann equation for the particle flux, $\phi_j(x, E)$, of ion of type j, with energy E, and depth x, as obtained from

$$\Omega \cdot \nabla \phi_j(x, \Omega, E) = \sum_k \int \sigma_{jk}(\Omega, \Omega', E, E') \phi_k(x, \Omega', E') dE' d\Omega' - \sigma_j(E) \phi_j(x, \Omega, E)$$

where σ_j is the total reaction cross section and σ_{jk} is the channel changing cross sections. The HZETRN code solves the Boltzmann equation using the continuous slowing down approximation. The straight-ahead approximation is used for projectile nuclei (Wilson et al., 1991) and angular dependence of scattering is considered for neutrons (Clowdsley et al., 2001). Details on the numerical methods used in this transport code can be found in refs. (Wilson et al., 1991) and (Clowdsley et al., 2001). Nuclear fragmentation cross sections are described by the quantum multiple scattering (QMSFRG) model (Cucinotta et al., 1998; Cucinotta et al., 1997). The QMSFRG model considers the energy dependence of the nucleus-nucleus interaction, quantum effects in nuclear abrasion, and a stochastic model of the de-excitation of pre-fragment nuclei produced in projectile-target nuclei interactions. The organ dose equivalent, H_T can be determined by integrating the particle energy spectra folded with the energy, mass, and charge dependent stopping power or linear energy transfer (LET), $L(E)$, and the LET dependent quality factor, $Q(L)$, and by considering the distribution of shielding at the tissue,

$$H_T = \sum_j \int dE \phi_j(E) L(E) Q[L(E)]$$

The organ dose equivalent, H_T is expressed in the units of Sievert (Sv) (1 cSv = 1 rem). Values of the LET dependent Q vary between 1 and 30 for the GCR with the highest values for LET = 50–200 keV/μm in the range of GCR ions with charge, $Z = 14$–26.

The modulation of the GCR near Earth is described using the Badhwar et al., (Badhwar et al., 1992; Badhwar et al., 1994) model in terms of the magnetic field deceleration parameter, Φ. GCR spectra for several solar maximum ($\Phi = 1060 - 1216$ MV) and solar minimum ($\Phi = 428$ MV) scenarios were generated. Differences in solar modulation and GCR composition near-Earth and near-Mars have not been considered in the calculations. Webber (Webber et al., 1987) has estimated an increase in the modulation potential Φ of 10 MV per A.U., which suggests the change in GCR between Earth (~ 1.0 AU) and Mars (~ 1.5 AU) would be small, however the modulation would be in the energy region where HZE particles have maximum biological quality factors and confirmation of the change due to modulation is needed for human exploration. The simultaneous observation of GCR near-Earth and in Mars orbit is one approach to investigate the effects of the radial gradient on GCR. Energy spectra for the six most abundant GCR elements are fit to satellite data (H, He, C, O, Si, and Fe). The other GCR ions are scaled using values in the literature as described previously (ref. (Badhwar et al., 1994) and references cited therein).

The number of particle-hits per cell nucleus due to direct particle traversals (Cucinotta et al., 1998) is evaluated using an average cell cross-sectional area of 100 μm^2. Calculations that include indirect cell hits from delta rays (secondary electrons) produced by ions will be reported elsewhere. Also, any additional shielding advantage from the Martian surface magnetic fields (assumed to be weak and sparse at this time) will be addressed and reported in future when such data is available.

Low and high-density Mars atmospheric models (16 and 22 g/cm^2), assuming a spherically distributed CO_2 atmosphere, are considered for all calculations (Simonsen et al., 2000). However, for simplicity, in this report only the high-density model (22 g/cm^2) calculations are presented. Atmospheric density is known to vary inversely with respect to the altitude from the mean surface. Based on the

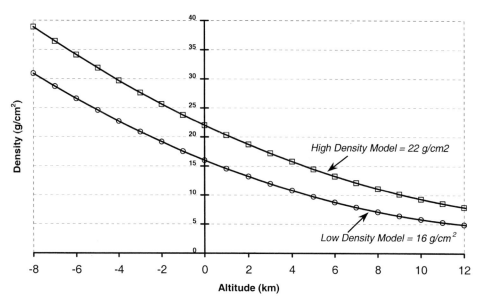

Figure 1. Density model of the CO_2 atmosphere on the Martian surface. The two density models, low-density (16 g/cm^2) and high-density (22 g/cm^2) are generally considered at the mean surface (0 km). The variation in the CO_2 density at a given altitude (-8 to $+12$ km) is derived, as shown in this figure, for both the density models from the NASA published data (Simonsen, 1997). All particle flux calculations presented in this report are only for the high-density (22 g/cm^2) CO_2 model with altitude variation.

NASA published data (Simonsen, 1997), we have derived a model to assess the variation in the atmosphere with respect to the altitude. Figure 1 plots CO_2 density versus altitude on the Mars surface for the low- and high-density models. Figure 2 describes the geometry used in the calculations. Seasonal variations of atmospheric density are not included in these calculations and hence we refer to as the static atmospheric density model in our calculations.

The resultant shielding offered by the CO_2 atmosphere at a given altitude location is calculated for a set of geodesic distributed rays using the relation

$$s(z, \theta) = \sqrt{(R+h)^2 \cos^2(\theta) + [2R(z-h) + z^2 - h^2]} - (R+h)\cos(\theta)$$

where, h is the altitude above the mean surface, s being the distance along the slant path with zenith angle, θ (is calculated between 0 and 90^0 with 1^0 increments) and z is the vertical length of the atmosphere above the identified location. Radiation transport along the rays are then evaluated using the HZETRN code and results integrated over z and h to obtain particle flux at each location on the Martian surface.

Calculations of radiation transport in the human body, which locate astronauts at specific surface locations on Mars, were performed using the methods described

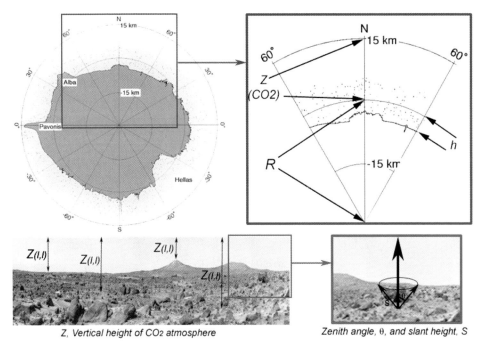

Figure 2. Top left image is an illustration of the global profile of the Martian surface from data obtained by MOLA crossing longitudes 52° E and 247° E. The north and south polar caps are at the top and bottom with a vertical exaggeration of 100:1 (adopted from Smith *et al.* 1999). Top right is the description of the mean surface radius, R, vertical height, Z, of the CO_2 atmosphere above the mean surface, and height of the target point above / below the mean surface as used in the calculations. Bottom left is the Martian surface as seen from the Pathfinder landing site with the description of variations in the vertical height of the Martian atmosphere for a given longitude and latitude values, Z(l,l). Bottom right is the description of the slant height, *s*, and the zenith angle, used in the calculations for the effective shielding from the atmosphere at a given target point on the Martian surface.

in conjunction with computerized anatomical models (Billings *et al.*, 1973) to represent the self-shielding of the human body. Particle energy spectra, organ dose equivalent, and cell hits were evaluated at 12 representative different anatomical locations for solar minimum and solar maximum. Visualization of the Martian surface data was accomplished using the computer routines (VIZ-MARS) that we have developed for this application using the software package *MATLAB*.

3. Results and Discussion

By combining the organ dose equivalent with age and gender specific risk coefficients, an estimate of the probability of fatal cancer from radiation exposure is made (National Academy of Sciences, 1997; Cucinotta *et al.*, 2001). We note that the HZETRN model has been shown to predict the GCR dose equivalent to within

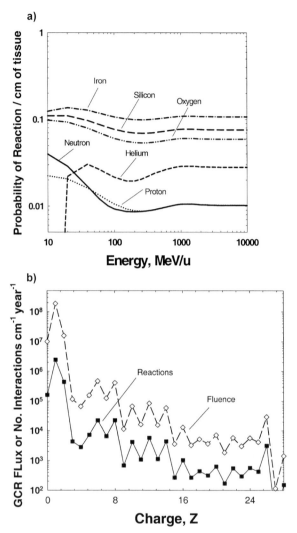

Figure 3. (a) The probability of a nuclear reaction per cm of tissue traversed versus energy for several ions prominent in the GCR; (b) Comparisons of the elemental fluence of GCR ions near solar minimum to the number of nuclear interactions that occur per cm of tissue traversed by each charge GCR group. Calculations are made behind 5 cm of tissue.

±25% accuracy. However, a much larger uncertainty exists in the understanding of the radiobiology of radiation effects at low dose-rates and of cancer risks from heavy ions (National Academy of Sciences, 1997; Cucinotta et al., 2001). Estimates of cell damage (Cucinotta et al., 1995; Cucinotta et al., 2000) have been considered along with the current organ dose equivalent to illustrate the role of attenuation of heavy ions by shielding such as the Martian atmosphere, spacecraft, and tissue shielding.

TABLE I

The effectiveness of water shielding (5 g/cm^2 and 10 g/cm^2) in transit form Earth to Mars on the organ dose equivalent at 12 different locations in the human body during solar minimum condition. The 'Point' dose refers to dose without body shielding and the 'Skin' organ location refers to the average dose in the human body at the skin level.

Organ	Organ Dose Equivalent (cSv/yr)				
	0 cm	5 cm	10 cm	%Reduction by 5 cm	%Reduction by 10 cm
Point	120.0	95.6	74.3	20.3%	38.1%
Skin	94.1	73.2	60.0	22.3	36.2
Eye	96.0	73.7	60.5	23.2	37.0
BFO	70.0	58.2	50.7	16.9	27.6
Bladder	60.1	52.0	46.7	13.5	22.4
Colon	69.0	57.7	50.5	16.4	26.9
Esophagus	65.9	55.6	48.9	15.6	25.7
Gonads	73.7	60.0	51.7	18.6	29.9
Liver	63.5	54.1	48.0	14.8	24.4
Lung	66.8	56.2	49.4	15.9	26.1
Stomach	61.0	52.6	47.1	13.7	22.7
Thyroid	72.6	59.8	51.8	17.6	28.7

The major physical processes that modulate the heavy ions are atomic energy loss processes that are well described by stopping powers, and nuclear reactions including fragmentation and production processes (Wilson *et al.*, 1991; Cucinotta *et al.*, 1998; Zeitlin *et al.*, 1997). Figure 3a shows calculations of the probability that a GCR ion will suffer a nuclear collision per cm of path-length in water. Figure 3b shows the elemental distribution of the GCR and calculations of the number of nuclear reactions made by each GCR charge group per path-length in water. These results illustrate that a large number of interactions of the GCR will occur with both tissue and the Mars atmosphere. However, the level of shielding provided is insufficient to completely eliminate the heavy ion components and also an accurate description of both fragmentation and transport is needed.

In Figure 4 we show results for the number of particle hits per cell as a function of altitude on the Mars surface for distinct charge groups at solar minimum and solar maximum conditions. The number of cell hits by light ions such as protons and helium are only modestly affected by atmospheric shielding (altitude variations). This is due to the changes in the balance of loss from atomic slowing down processes and gain from the fragmentation of heavy ions or production from

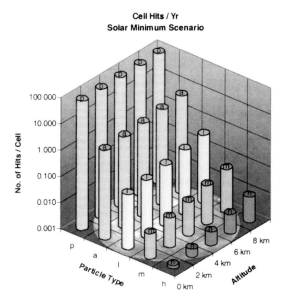
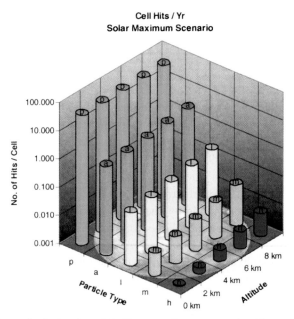

Figure 4. Comparison of calculated particle hits per cell per year at the skin on the Martian surface for solar minium and solar maximum conditions. Calculations include the average body shielding on the skin for the 50% percentile male Billings et al. 1973. Results for protons, alpha, light, medium, and heavy charge groups are shown.

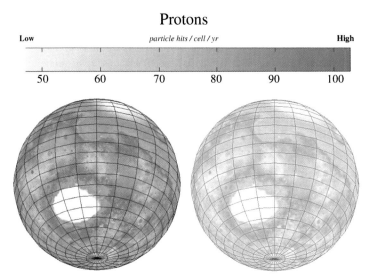

Figure 5. Estimates of the number of proton hits per cell per year on the Martian surface. Calculations consider the transport of the GCR through the Mars atmosphere using the MOLA topographical data and include the average body shielding on the skin for the 50th percentile male (Billings et al., 1973). Calculations consider the extreme solar cycle scenarios with calculations with left panel near solar minimum (with solar deceleration parameter, = 428 MV) and right panel solar maximum scenario (= 1050 MV) is shown.

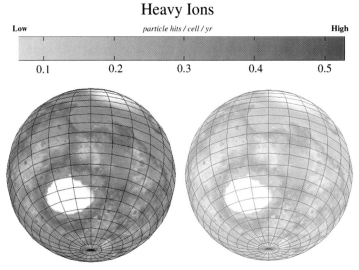

Figure 6. Estimates of the number of heavy ion ($Z > 2$) hits per cell per year on the Martian surface. Calculations consider the transport of the GCR through the Mars atmosphere using the MOLA topographical data include the average body shielding on the skin for the 50% percentile male (Billings et al., 1973). Calculations consider the extreme solar cycle scenarios with calculations with left panel near solar minimum (with solar deceleration parameter, = 428 MV) and right panel solar maximum scenario (= 1050 MV) is shown.

TABLE II

The effectiveness of water shielding (5 g/cm^2 and 10 g/cm^2) on the Mars surface for the organ dose equivalent at 12 different locations in the human body during solar minimum condition. The 'Point' dose refers to the dose without body shielding and the 'Skin' organ location refers to the average dose in the human body at the skin level.

Organ	Organ Dose Equivalent (cSv/yr)				
	0 cm	5 cm	10 cm	%Reduction by 5 cm	%Reduction by 10 cm
Point	25.8	21.3	20.1	17.4%	22.1%
Skin	19.3	18.6	18.0	3.8	6.8
Eye	19.7	18.8	18.2	4.4	7.7
BFO	19.4	18.7	18.0	3.9	6.9
Bladder	21.1	19.7	18.9	6.4	10.5
Colon	19.2	18.5	18.0	3.6	6.5
Esophagus	19.1	18.5	17.9	3.4	6.2
Gonads	19.8	18.9	18.3	4.3	7.5
Liver	19.6	18.8	18.2	4.0	7.2
Lung	19.6	18.8	18.2	4.0	7.1
Stomach	21.1	19.8	19.0	6.1	10.2
Thyroid	19.0	18.3	17.8	3.4	6.1

TABLE III

Percent excess fatal cancer risk projections and 95% C. I.'s for 40-year-old females or males. Calculations are for opposition or conjunction type Mars missions using 4 g/cm^2 aluminum shielding and high density Mars CO_2 atmosphere and considering effects of the addition of 10 cm water shielding. Values in parenthesis indicate days on Mars surface Cucinotta et al., 2001.

Mission Type	Days (Mars or lunar surface)	0 cm H$_2$O	10 cm H$_2$O
	Females		
Opposition	360 (30)	3.3 [0, 18.0]	2.5 [0, 14.6]
Opposition	660 (30)	6.2 [0, 34.0]	4.6 [0, 27.5]
Conjunction	1000 (600)	5.7 [0, 30.8]	4.5 [0, 25.6]
	Males		
Opposition	360 (30)	2.0 [0, 10.8]	1.5 [0, 8.8]
Opposition	660 (30)	3.7 [0, 20.4]	2.8 [0, 16.5]
Conjunction	1000 (600)	3.4 [0, 18.5]	2.7 [0, 15.3]

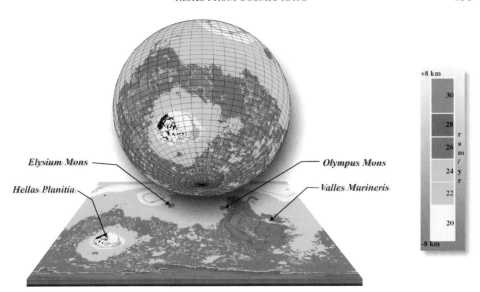

Figure 7. Calculations of the skin dose equivalent for astronauts on the surface of Mars near solar minimum. The variation in the dose with respect to altitude is shown. Higher altitudes (such as Olympus Mons) offer less shielding from the CO_2 atmosphere and lower altitudes (such as Hellas Planatia). The effective total dose has a range between 20 and 30 cSv/yr as a function of altitude for the static atmospheric high-density CO_2 model used here.

atoms in the atmosphere. In contrast, heavy ions undergo a large reduction with increasing atmosphere (decreasing altitude). In the present model we also predict that a significant contribution of heavy ion component occurs, especially at higher altitudes. Neutrons (not shown here) are less sensitive to changes in the amount of atmosphere since they contain an important back-scattered component at lower energies (< 100 MeV) (Clowdsley *et al.*, 2001). Because neutrons will make up a larger contribution at low altitudes and production cross sections for neutrons are sensitive to target mass, information on atmospheric and soil composition will be important for designing surface habitats for astronauts. The 2001 *Mars Odyssey* spacecraft [References this special issue] and future Lander missions should provide such data. Figure 5 and Figure 6 show results for the Martian radiation climate over the surface of Mars for protons and heavy ions respectively with both solar minimum and maximum conditions. These results indicate significant reduction for heavy and medium ions, moderate reduction for light ions, and little or no reduction for protons and neutrons (neutrons are not shown). Because of the large uncertainties in the biological effects of heavy ions, validation of these results by measurements on the surface of Mars will be essential for future human safety.

Table I compares the organ dose equivalent en-route to Mars and Table II on the Mars surface for the major tissues that contribute to radiation cancer risks during a solar minimum condition. Included in these calculations are results that consider the addition of shielding configurations of 5 or 10 cm of water equivalent material. The cruise phase of the mission could lead to organ doses that approach or exceed legal exposure limits used in low Earth orbit (LEO), which correspond to an excess probability of 3% fatal cancer (National Council on Radiation Protection and Measurements, 2000). Dose limits are age and gender dependent as determined by epidemiology data. For manned exploration of Mars two types of missions are considered; conjunction-class where the mid-point of the mission is closer to Mars (Mars is on the same side of the Earth from Sun) and opposition-class where the mid-point is further from Mars (Mars is on the opposite side of the Earth from Sun). Conjunction-class missions involve longer stay on Mars (300–600 days) with one-way transit times of 150-250 days. Opposition-class missions involve shorter stay on Mars (20–60 days) with one-way transit times of 100–400 days (Cucinotta *et al.*, 2002). The opposition-class missions require higher energy requirements. Table III shows calculations of fatal cancer probabilities for 40-year-old males and females for such scenarios (Cucinotta *et al.*, 2001). The 95% confidence levels shown are based on calculations, which consider the uncertainties in epidemiological data, radiation qualities, and the understanding of the physical environment. In Table III the comparison with the addition of 10 cm water-shielding shows that the addition and optimization of shielding could lower risks to a level within the acceptable level of risk for LEO astronauts. However, the large uncertainties limit any conclusion on the acceptability of risk for long missions (> 100 d). The comparison in Table III shows that maximizing crew times on the Mars surface (conjunction-class missions) because of the lower surface doses will be favorable for human exploration. In Figure 7 we show a climatic map of the skin dose equivalent for the entire surface of Mars. Variations on the order of 50% are predicted due to changes in atmosphere. Although, uncertainties in heavy ion radiobiology including the radiation quality factors are expected to be much larger at this time than estimates of physical quantities, a validation of the radiation climate map described here is needed. Surface validation measurements should include measurements of LET spectra and the integral dose and dose equivalent (References this special issue) along with spectroscopic data on neutrons.

4. Conclusions

Calculations of the Martian radiation climate illustrate the level of detail that is now available by the most recent computer codes developed by NASA. Critical questions to be addressed include what is the current accuracy of these calculations, what accuracy will be required, and what measurements are needed to validate these models? Previous comparisons with measurements near Earth provide some validation of the models used here and an agreement to within 25% is found for the dose equivalent. However, clearly data near Mars and most importantly on the surface of Mars will be needed in the context of the high level concern of biological risks to astronauts from heavy ions. At present, NASA Johnson Space Center's MARIE (Martian Radiation Environment Experiment) instrument on board the *2001 Mars Odyssey* spacecraft has been collecting and successfully providing data for the Martian radiation environment mapping since March 13, 2002 (*http://srhp.jsc.nasa.gov/*; *http://marie.jsc.nasa.gov/*). In near future, other radiation instruments are being planned to land on Martian surface and provide similar data. Once these measurements are available, they will be utilized to verify our calculated predictions and update the models as needed. Local measurements on Martian surface including atmospheric density variations and regolith contributed albedo radiation will be valuable for designing future human exploration missions to Mars.

This work is dedicated to the memory of Dr. Guatam D. Badhwar (1940-2001) of NASA-JSC.

References

Badhwar, G., *Space Science Reviews* **110**, 131–142.
Badhwar, G. D., and O'Neill, P. M.: 1992, 'An Improved Model of GCR for Space Exploration Missions', *Nucl. Tracks Radiat. Meas.* **20**, 403–410.
Badhwar, G. D., Cucinotta, F. A., and O'Neill, P. M.: 1994, 'An Analysis of Interplanetary Space Radiation Exposure for Various Solar Cycles', *Radiat. Res.* **138**, 201–208.
Billings, M. P., Yucker, W. R., and Heckman, B. R.: 1973, Body Self-Shielding Data Analysis, McDonald Douglas Astronautics Company West, MDC-**G4131**.
Clowdsley, M. S., Wilson, J. W., Kim, M., Singleterry, R. C., Tripathi, R. K., Heinbockel, J. H., Badavi, F. F., and Shinn, J. L.: 2001, 'Neutron Environments on the Martian Surface', *Physica Medica* **17**, 94–96.
Cucinotta, F. A., and Dicello, J. F.: 2000, 'On the Development of Biophysical Models for Space Radiation Risk Assessment', *Adv. Space. Res.* **25**, 2131–2140.
Cucinotta, F. A., Wilson, J. W., Katz, R., Atwell, W., and Badhwar, G. D.: 1995, 'Track Structure and Radiation Transport Models for Space Radiobiology Studies', *Adv. in Space Res.* **18**, 183, 194–203.
Cucinotta, F. A., Wilson, J. W., Shinn, J. L., Tripathi, R. K., Maung, K. M., Badavi, F. F., Katz, R., and Dubey, R. D.: 1997, Computational Procedures and Data-Base Development. In *NASA Workshop on Shielding Strategies for Human Space Exploration*. Eds. Wilson, J. W., Miller, J., Konradi, A., and Cucinotta, F. A., NASA CP-**3360**.
Cucinotta F. A., Nikjoo H., and Goodhead D. T.: 1998, 'The Effects of Delta Rays on The Number of Particle-Track Traversals per Cell in Laboratory and Space Exposures', *Radiat. Res.* **150**, 115–119.
Cucinotta, F. A., Wilson, J. W., Tripathi, R. K.; and Townsend, L. W.: 1998, 'Microscopic, Fragmentation Model For Galactic Cosmic Ray Studies', *Adv. in Space Res.* **22**, 533–537.
Cucinotta, F. A., Manuel, F. K., Jones, J., Izsard, G., Murray, J., Djojenegoro, and Wear, M.: 2001, 'Space Radiation and Cataracts in Astronauts', *Radiat. Res.* **156**, 460–466.
Cucinotta, F. A., Schimmerling, W., Wilson, J. W., Peterson, L. E., Badhwar, G. D., Saganti, P. B., and Dicello, J. F.: 2001, 'Space Radiation Cancer Risks and Uncertainties for Mars Missions', *Radiat. Res.* **156**, 682–688.
Cucinotta, F. A., Badhwar, G. D., Saganti, P. B., Schimmerling, W., Wilson, J. W., Peterson, L., and Dicello, J.: 2002, Space Radiation Cancer Risk Projections for Exploration Missions: Uncertainty Reduction and Mitigation, NASA TP-**21077**.

National Academy of Sciences Space Science Board, HZE Particle Effects in Manned Space Flight, National Academy of Sciences U.S.A. Washington D.C., 1973.

National Academy of Sciences, NAS. National Academy of Sciences Space Science Board, Report of the Task Group on the Biological Effects of Space Radiation. Radiation Hazards to Crews on Interplanetary Mission National Academy of Sciences, Washington, D.C., 1997.

National Council on Radiation Protection and Measurements, Radiation Protection Guidance for Activities in Low Earth Orbit, NCRP Report **132**, Bethesda MD, 2000.

Simonsen L. C.: 1997, Analysis of Lunar and Mars Habitation Modules for the Space Exploration Initiative, Chapter-4 in *Shielding Strategies for Human Space Exploration*, Ed. J. W. Wilson, J. Miller, A. Konradi, and F. A. Cucinotta, NASA CP-**3360**, 43–77.

Simonsen, L. C., Wilson, J. W., Kim, M. H., and Cucinotta, F. A.: 2000, 'Radiation Exposure for Human Mars Exploration', *Health Phys.* **79**, 515–525.

Smith, D. E., Zuber, M. T., Solomon, S. C., Philips, R. J., Head, J. W., Garvin, J. B., *et al.*: 1999, 'The Global Topography of Mars and Implications for Surface Evolution', *Science* **284**, 1495–1503.

Webber, W. R.: 1987, 'The Interstellar Cosmic Ray Spectrum and Energy Density. Interplanetary Cosmic Ray Gradients and a New Estimate of the Boundary of the Heliosphere', *Astron. Astrophys.* **179**, 277–284.

Wilson J. W., Townsend, L. W., Schimmerling, W., Khandelwal G. S., Khan, F., Nealy, J. E., Cucinotta, F. A., Simonsen, L. C., and Norbury, J. W.: 1991, Transport methods and interactions for space radiations, NASA-RP**1257**.

Zeitlin, C., Heilbronn, L., Miller, J., Rademacher, S. E., Borak, T., Carter, T. R., Frankel, K. A., Schimmerling, W., and Stronach. C. E.: 1997, 'Heavy Fragment Production Cross Sections for 1.05 GeV/nucleon ^{56}Fe in C, AL, Cu, Pb, and CH_2 Targets', *Phys. Rev. C* **56**, 388–397.

In Memoriam, Gautam D. Badhwar (1940–2001)

Gautam D. Badhwar, Principal Investigator and Chief Scientist for Space Radiation at the NASA Johnson Space Center and a contributor to this volume, was born on November 8, 1940 near Bombay, India and died on August 28, 2001 in Houston, Texas

Gautam worked at the Tata Institute of Fundamental Research in Bombay while completing his undergraduate degrees. In 1967 he received his Ph.D. in Physics from the University of Rochester NY, where he worked for several years as a research associate. In 1972 as a National Academy of Science Fellow, he came to Johnson Space Center (JSC) where he accepted a permanent position in 1974.

Gautam's early work in India and at the University of Rochester pioneered the measurement of cosmic rays, including the study of the nuclear interactions of cosmic rays, the interplanetary electron and gamma-ray compositions, and many firsts in the measurement of energy spectra and isotopes of hydrogen, helium ions and heavy ions. Gautam liked to master the instruments with which he worked, and performed pioneering studies of the light response of scintillators and of the energy loss processes in silicon detectors. His work was known worldwide by the early 1970s and was influential in understanding characteristics of the particle beams used in the treatment of cancer patients at Lawrence Berkeley Laboratory in the 1980s and more recently at hospitals in Germany and Japan.

After arriving at Johnson Space Center in 1972, Gautam made major contributions to NASA's manned space program and both Earth and interplanetary science goals. His earliest work at JSC included the first measurements of anti-matter in space. In the early 1980s Gautam made valuable contributions to the Earth Observation Program at Johnson Space Center. Over the years Gautam-designed instruments have flown on many satellites and spacecraft including OSO, the Space Shuttle, the Russian space station, Mir, and more recently the International Space Station (ISS) and the MARIE experiment on the Mars Odyssey spacecraft. These instruments will orbit the Earth and Mars for many years to come providing us with valuable data.

Gautam's invaluable contributions focused on exposures to radiation risk in space and thus supported the health and safety of NASA's astronauts. His energy and enthusiasm turned instruments like the Tissue-Equivalent Proportional Counter (TEPC) into reality. The careful measurements made by Gautam and his colleagues determined the quality of the radiation to which astronauts on Shuttle, Mir and the International Space Station are exposed. Gautam and his colleagues' models of galactic cosmic rays and Earth's radiation belts are vital for projecting exposures that astronauts will receive in future manned space flights. In 2001 Gautam co-authored a study of the uncertainties for projecting radiation risks for Mars' missions. This study shows that, of all the contributions to the uncertainties in risk projections, the environmental ones are now the smallest. This is a true measure of the achievements provided to NASA over the course of Gautam's career. This work will be used by other scientists for planning future exploration missions, making the task of preparing for Mars' exploration much easier.

More recently Gautam had worked on measuring the effects of potential shielding materials on the attenuation of space radiation. These measurements have validated shielding approaches that will be used in the crew quarters of the ISS. Other work studied the Earth's trapped radiation belts and their evolution with time. His instruments provide a warning alarm for astronauts and cosmonauts on the ISS to protect themselves from high radiation levels caused by solar storms. It will be difficult to find someone able to continue his work at the same level of competence, and even more difficult to find someone to provide the leadership he offered.

Gautam served as a member of the Challenger accident investigation team and contributed to the understanding of the beneficial effects of radiation on crop growth, the effects of radiation on electronics damage, studies of cold fusion, the origin and composition of stars and in many other areas that would take too long to note here. He exemplified the spirit of international cooperation that made the NASA-Mir Phase I program a success. This work brought about a tremendously rewarding partnership between NASA and the Russian Space Agency that has lasted for nearly a decade. His experiments testing spaceflight components at particle accelerators throughout the world brought about many friendships and collaborations that increased the value of his work to NASA and the scientific community.

Gautam was a very devout Hindu. One of the teachings of Hinduism is that life can be viewed as an illusion and man must struggle to break through the illusions of the world to achieve the reality of one's own life. Many theoretical physicists have utilized the results of Gautam's numerous experimental findings to validate their models. It is truly fitting that because of Gautam's work a merging of the illusions of abstract theoretical concepts and the electronic signals from detectors came together time and time again to achieve a single reality as represented by the agreement between the two worlds of experimental and theoretical physics.

Gautam has received many awards throughout his career including the Tuli Gold Medal in 1959, an Eastman Kodak Prize for Research in 1965, numerous NASA awards for superior achievement, NASA Exceptional Scientific Achievement Medals in 1991 and 1992, NASA award for Support of the STS-51L Challenger Accident, and a Spaceflight Awareness Award in 2001. He co-authored over 200 refereed journal publications during his career and served as a member of several committees of the National Council on Radiation Protection and Measurements (NCRP) and on the Committee on Space Research (COSPAR).

Gautam exemplified leadership at several levels including his breakthrough approach to science, his ability to establish and maintain collaborations, and his many committee memberships. He also was able to guide his support team to sustain a 110% effort over many years, which brought great rewards to themselves and NASA. He mentored many students at the undergraduate and graduate levels and young scientists in post-doc and other positions. He was greatly admired for the effective way he worked within the NASA system.

Throughout the years Gautam continued to be an example of elegant simplicity in his work and in his life, willing at all times to share his knowledge and his friendship with his colleagues. His sterling human qualities make this a personal loss and a source of great sadness. Gautam is dearly missed by all of us who were lucky to be his friend.

F. A. Cucinotta